A FALCON FIELD GUIDE™

Rocks, Gems, and Minerals

Garret Romaine

FALCONGUIDES

GUILFORD, CONNECTICUT
HELENA, MONTANA
AN IMPRINT OF GLOBE PEQUOT PRESS

To my uncle Doug, who never turned down
a chance to head out to the field.

To buy books in quantity for corporate use
or incentives, call **(800) 962-0973**
or e-mail **premiums@GlobePequot.com**.

FALCONGUIDES®

Interior photos by Garret Romaine except where otherwise credited.

Project editor: Heather Santiago
Text design: Sheryl P. Kober
Layout: Sue Murray

Library of Congress Cataloging-in-Publication Data is available on file.

ISBN 978-0-7627-8042-6

Printed in the United States of America

10 9 8 7 6 5 4 3 2 1

Contents

Acknowledgments

I owe a lot of people a huge debt of gratitude. At Portland State University, Dr. Martin J. Streck pulled some amazing specimens from his mineralogy lab so I could show you a real hornfels and a true graywacke. Tim Fisher, producer of the *Ore-Rock-On* DVD that pinpoints 2,700 collecting locales throughout the Pacific Northwest, has been a valuable resource and friend over the years. Lanny Ream, veteran field geologist and author of several excellent books on local rockhounding and field collecting, has proved to be a great resource, as has Bob Jackson, owner of Geology Adventures, the Spruce Ridge mine, and the Rock Candy Mine. Rudy Tschernich, former curator of the Rice Museum in Hillsboro, Oregon, and author of *Zeolites of the World,* has been a huge help over the years. Martin Schippers, Dirk Williams, Frank Higgins, Val Bailey, Terry Snider, and Jake Riley have been particularly helpful on recent expeditions to explore this great country.

On the production side, Kleur Pre-Media Services in Lynnwood, Washington, tirelessly tuned up the photographs for me. Rachel Houghton helped with the photography and the manuscripts. Numerous rock shops and museums allowed me to photograph their specimens that were better than mine. I owe a special thanks to Ed's House of Gems and Jim Smith at Earthly Treasures for letting me bring my camera to their shops.

I especially want to thank Melena Wallace, executive director of the famed Rice Northwest Museum of Rocks and Minerals in Hillsboro, Oregon, for access to their fantastic collection, with numerous specimens gracing these pages.

Finally, my family deserves thanks. My uncle Doug Romaine has been a wonderful advisor over the years, encouraging me to first seek a geology degree and then leading me on adventures all over the West. This book is dedicated to him. And special thanks to my wife, Cindy, and the kids for putting up with a lifetime of talking about rocks.

Introduction

Goal of This Guide

The goal of this guide is to get you started from a casual observer to a competent rockhound, able to identify the most common rocks and minerals around you. In the ranking of this profession, there are many stages (left to right, increasing expertise):

Casual Observer	Pebble Pup	Rockhound	Prospector	Field Geologist	Mining Engineer

With this book you could conceivably make Prospector rank, able to at least classify just about anything, and know enough to be on the lookout for the rare and unusual. You need a lot more book learnin' to advance beyond capable prospector, able to direct the work of others. The vocabulary lesson alone that underpins geology is daunting. This book is intended as a gateway to collecting wonderful samples and improving your ability to identify what you see around you. But avoid the "get-rich-quick" attitude of a Yukon stampeder. Concentrate on building your own personal specimen cabinet, well stocked with samples you gather and augmented by what you have to buy for completeness. If you stumble upon a diamond mine or a ledge of emeralds, at least share the pictures. But don't expect it. Have fun, be safe, and don't quit your day job first.

About Geology

The term *geology* is a combination of two Greek words: *Geo* refers to the earth, and *logos* refers to the logic and language used to explain things. So think of geology as a way to organize and talk about the earth's processes. There are many things we can only guess at, so we base our theories on lab experiments and inferences that take many detailed drawings to explain and a lifetime to understand. Fortunately the more you see, the better things fall into place.

Geology is a young science, dating to 1815 if you start with William Smith's first geology map. Ancient Greeks, such as Pliny the Elder, had described minerals and metals, and many early scholars documented the metal mines of the times. Some of the smartest and most educated scientists laid down the basics, such as Nicolas Steno (1638–56), who observed that most of the time, the rocks at the bottom of a cliff are older than the rocks at the top. He called that the Law of Superposition, and it helped explain how fossil seashells ended up on mountaintops. He also developed the Principle of Original Horizontality: Sedimentary rocks are usually deposited flat, although there can be local pinching, advancing, and other variations. James Hutton presented his *Theory of the Earth* in 1785, Smith published a geology map of England in 1815, and then in 1830 Sir Charles Lyell wrote *Principles of Geology*. Great thinkers argued over whether geology happened in slow, methodical processes or in short, catastrophic bursts. Once these learned philosophers learned to measure the astonishing age of the earth—4.6 billion years—they realized that both arguments were true.

There are two key points to consider in geology: time and entropy.

1. **Time.** Earth is a very young planet and thus still very active. It's also pretty old, so it's all relative. Given enough time, a lot can happen on a young, geologically active planet. We have earthquakes, volcanoes, and moving continents, all stemming from the forces that boil up from the earth's magnetic core. Some activities happen quickly, like tsunamis. Other forces take millions of years, leaving only subtle clues, like all the mica flakes lined up in a schist. Good field observers can identify the obvious signs of things that seemed to happen before and apply those signs to the present and future.

2. **Entropy.** Things fall apart all the time. Stuff happens. Storms rearrange coastlines and rework river channels. Earthquakes, volcanoes, storms, and floods all move mountains and leave

scars that "heal." A rock balanced precariously atop another rock will not remain there for long; eventually it will shake loose. Earth is very efficient at recycling all that surface mayhem, hiding many clues. Mountains rise then get ground down under glaciers and unrelenting rain. Tight chemical bonds that hold atoms together eventually weaken thanks to water, heat, pressure, and time. Oxygen in the air constantly rusts iron and dissolves minerals. Those forces are always at work and easy to predict but sometimes hard to imagine.

Given enough time, almost anything can happen, and it usually does. We can only imagine what takes place at great depths. That's where the logic comes in. There is a lot of math, chemistry, physics, biology, and general science involved in sorting out what's going on in the field. But you're mostly interested in what you can see and collect, so read on.

Think in Series

We don't get a lot of absolutes in nature, so numbers, such as percentages of minerals present, help when thinking about crystal compositions. Just as there are probably no two snowflakes exactly alike, most granites differ in some way. You can't exactly tell without expensive equipment, but you can learn to eyeball specimens and compare them with known samples. And usually it doesn't really matter to that many decimals if you have a rough idea. You just want to collect the interesting forms, and you don't need a Ph.D. in structural geology to hammer out a seam of agate.

For another example, consider the series of variations between basalt and rhyolite. The range is completely understood but hard to identify:

Basalt	Andesite	Rhyolite
45%–55% Quartz	55%–65% Quartz	65%–75% Quartz
Lots of iron, aluminum, magnesium, and calcium	Intermediate accessories	Lots of potassium and sodium
Not very much potassium and sodium		Not very much iron, magnesium, and calcium

Minerals are definitely more straightforward than rocks, because they are more rigid in their chemistry. But even there, different concentrations of elements create a series; in some minerals the concentrations are referred to as impurities. Sodium chloride, known as halite or rock salt, is one sodium atom and one chlorine atom. There isn't a lot of wiggle room there. You can get some potassium in the crystal lattice, substituting itself for sodium, but that's an impurity. You can get a little boron once in a while where chlorine should be, and sometimes there is enough to haul out of the desert by mule team. Too many impurities and the result will more likely be a mineral with a different name.

Sedimentary rocks may seem uniform at first glance, but they are often composed of different ratios of sand, silt, and clay. So think more in terms of trending to one end of the scale or the other and less in absolutes.

Field Studies

One big complication in identifying rocks and minerals is the tendency of the earth's atmosphere to oxidize everything. That same oxygen we need to breathe also wreaks havoc on fresh material. Oxygen ions are always looking to hook up with another ion, preferably a metal. You've seen how a freshly sharpened knife blade starts to rust in the rain. That oxidation is also at work on cliff faces and boulders, aiding and abetting a tendency toward natural cracks, root action, freezing, and thawing. Many rocks get a reddish-brown surface in a short amount of time. First they turn color, and then they fall apart.

When you're trying to identify a mineral, you need to be able to see pure surfaces with visible crystal faces so that you can measure the angles and find patterns: cubic, hexagonal, or even rhombohedral. My personal favorite is the garnet, which has a dodecahedral crystal structure, meaning twelve-sided. You often need to break the rock to reveal fresh material and then look with a hand lens to see what the microcrystals can tell you. Every rockhound needs a hand lens!

Even then you can't always tell what you have. There are several traits to describe a rock or crystal, such as shape, size, streak, hardness, luster, color, and weight. I've provided a glossary at the end of this book to help you with many geological terms and expressions, and I've tried to keep the text as free as possible from the really daunting vocabulary.

HARDNESS

The hardness test is one of the most elementary ways to sort out samples. Developed by German mineralogist Frederich Mohs in 1812, it is still used today. It is not a linear scale. Try to keep a few of these common specimens on hand, and, if possible, memorize the ten standard listings. I included some of the "nonstandard" common test materials as well.

Mohs #	Standard Mineral	Common Test
1	Talc	
2	Gypsum	2½—fingernail
3	Calcite	3—wheat-back penny
4	Fluorite	
5	Apatite	5½—knife blade
6	Orthoclase feldspar	6–7—glass
7	Quartz	7+—hard steel file
8	Topaz	
9	Corundum	
10	Diamond	

Be a Kind Collector

Before going any further, let's talk about safety and etiquette in the field. The biggest safety tip is to never push your car beyond its limits, and bring tools. Upgrade your tires, and be religious about vehicle maintenance. Your expeditions can take you to the end of some very long and dangerous roads, so walking home won't be much fun.

Don't treat private property like public lands; get all the permissions you need beforehand or give up and move on. Obey all signs, control your kids and dogs, pick up litter, and take only what you need when collecting. Consider bringing along an extra bucket to carry out trash you may encounter when searching road cuts.

Armed with a hand lens, some common household items, and this book, you can start learning about the rocks and minerals you find out there. Good luck!

PART 1:

ROCKS

Igneous Rocks: Extrusive

The following table compares the various igneous extrusive rocks.

Matrix of Igneous Extrusive Rocks				
Name	Color	Grain Size	Composition	Useful Identification Characteristics
Pumice	Light	Fine	Sticky lava froth	Small bubbles
Andesite	Medium	Fine or mixed	Medium-silica lava	Plagioclase with pyroxenes
Felsite*	Light	Fine or mixed	Medium to high silica	Quartz
Dacite	Light	Mixed	Medium silica	Plagioclase with hornblende
Basalt	Dark	Fine or mixed	Low-silica lava	Has no quartz
Scoria*	Dark	Fine	Runny lava froth	Large bubbles
Rhyolite	Light to medium	Fine or mixed	Very high silica	Quartz is common
Obsidian	Dark	Fine	High silica	Glassy
Tuff	Light	Mixed	Andesite, rhyolite, or basalt	Fragments; sometimes welded

*= not covered in this guide

Pumice

This dacitic pumice is from the eruption of Mount St. Helens in 1980.

Group: Igneous; Extrusive
Key test(s): Floats; soft
Likely locale(s): Volcanic mountain ranges

Look for pumice in known volcanic regions, especially any area with recent events. Pumice is light enough to float, thanks to its frothy, glassy texture that traps air and makes each rock buoyant. Pumice has a unique texture and feel that makes it easy to identify, and it can show dramatic flow lines that further help. Many violent pyroclastic events spew pumice along with ash and other material, causing pumice fragments to show up in breccias, tuffs, and other ash deposits. Rivers that drain volcanic regions, such as the Columbia River in the Pacific Northwest, are often lined with small pumice fragments. Pumice rafts are also common after volcanic eruptions in a marine environment. Collectors don't need more than a sample or two for a complete collection unless considering industrial uses for pumice, such as yard rock, potting soil, and abrasives.

Andesite

This andesite specimen is from the Cascade Mountains, Oregon.

PHOTO COURTESY OF RICE NORTHWEST MUSEUM OF ROCKS AND MINERALS

Group: Igneous; Extrusive
Key test(s): Fine grained with pyroxene and plagioclase
Likely locale(s): Volcanic mountain ranges

Andesite is a common form of lava that dominates many volcanic mountain ranges, as its higher silica content causes it to pile up better as opposed to flood basalts, which flow for miles. For example, Mount Rainier in Washington is about 90 percent andesite. This rock is named for the Andes Mountains, where it is a key component of those volcanoes. Andesite often exhibits a platy structure, appearing in outcrops as though it was stacked, but that isn't a telltale clue. Many lavas show flow lines and horizontal fracturing. Andesite is the extrusive equivalent of diorite, matching that rock in chemical composition. This lava is typically fine grained, with small crystals of plagioclase feldspar (such as andesine), hornblende, pyroxene (such as augite or diopside), and biotite. It makes a fine decorative rock, weathering from its light gray appearance when fresh to a dark gray or black color. Andesite sometimes displays cavities and pockets, but it is not as well known for opal or agate blebs. However, it is sometimes encrusted with secondary minerals such as epidote, making for excellent display pieces.

Dacite

This dacite is from Mount St. Helens, Washington. Inset is Mount St. Helens dacite dome.

SPECIMEN PHOTO COURTESY OF DAVID TUCKER, WESTERN WASHINGTON STATE UNIVERSITY

Group: Igneous; Extrusive
Key test(s): Hornblende and plagioclase in fine-grained matrix
Likely locale(s): Volcanic ranges; often in dikes or sills

In composition, dacite ranks between andesite and rhyolite, typically featuring much larger crystals and a coarse, patchy appearance. Dacite is not common, so it can be difficult to get familiar with it. The color is quite variable, ranging from white, gray, and rarely black to pale red or brown or even to deeper reds and browns. The matrix frequently appears oxidized. Mineralogically, dacite contains the same components as granodiorite, with plagioclase feldspar, quartz, biotite, hornblende, and augite. Dacite is usually found as ignimbrite or tuff rather than in a flow, but it can display some flow banding. One famous deposit of dacite is the growing lava dome in the exploded crater of Mount St. Helens, which is serving as a plug atop the active volcano.

Basalt

This columnar basalt is from the Middle Fork of the John Day River, Oregon.

Group: Igneous; Extrusive
Key test(s): Fine grained, often with microscopic crystals only; can form dramatic columns
Likely locale(s): Volcanic regions

Basalt is common in many parts of North America. It appears gray or light black when fresh but quickly weathers to a tan, yellow, or brown surface. It can be difficult to identify because it tends to be fine grained, so look for clues in the deposit itself. Vesicles are common; these holes sometimes fill with agate, chalcedony, or opal in small, round blebs that often weather out and are collectible. Bubbly basalt is called scoria and usually indicates the top of the flow where there is considerable froth. When larger crystals are identifiable, the rock is termed a porphyry. While basalt itself is rarely of interest to rockhounds, under the right conditions it can host zeolites, quartz veins, and calcite veins. When erupted underwater, basalt forms "pillows" and can display a yellow or orange crust of glassy material called palagonite. Look for evidence of low viscosity as the basalt flow moved—andesite and rhyolite are more noted as viscous lava that builds up faster, while basalt can flow hundreds of miles. Although considered extrusive, basalt can form in large sills and cool slowly enough to create impressive polygons, referred to as columnar jointing. These can form impressive colonnades and are collectible as impressive towers.

5

Rhyolite

Yellowstone Canyon in Wyoming is one of the most famous rhyolite exposures in the world.

Group: Igneous; Extrusive
Key test(s): Quartz crystals in fine-grained matrix
Likely locale(s): Pyroclastic flows and caldera fills

Rhyolite is another extremely common igneous rock, so learning to distinguish it from andesite and basalt is important. Rhyolite's primary minerals include quartz, feldspar, biotite, and hornblende, with accessory magnetite. Its high silica content usually means rhyolite is lighter in color. Rhyolite is usually gray but can appear yellow, pale yellow, and from pale to deeper red. Rhyolite cools faster than basalt and usually appears to be composed of fine-grained but distinguishable crystals. It is sometimes glassy. Flow banding is common, and crystals often show alignment under a hand lens, as do any small vesicles present. Like basalt, rhyolite is common throughout North America; probably the most famous exposure is in the Yellowstone Canyon, pictured above.

Obsidian

"Snowflake" obsidian from northern Nevada is a rare form of this volcanic glass.

Group: Igneous; Extrusive
Key test(s): Glassy luster; sharp edges
Likely locale(s): In rhyolite flows

Obsidian has no crystal structure—it cooled too fast for the atoms to align in a crystal lattice—so it is not a mineral; technically it is a glass. It's an easy rock to identify because it has a glassy luster when freshly chipped. However, when weathered or rolled in streams and rivers, with no fresh surface, obsidian can resemble rhyolite pebbles. Other than the classic shiny black luster, look for evidence of flow, such as banding. The familiar fracture pattern is called conchoidal, or shell-like. Knappers have long used obsidian's tendency to break predictably to fashion amazing stone tools. The most common obsidian is jet-black, but there are other colors and forms. At famed Glass Buttes in southeast Oregon, there are green, brown, root beer, banded, rainbow, and other varieties. At Davis Creek in northern California, obsidian forms in long needles. Another form of obsidian is known as Apache tears.

Tuff

Volcanic ash, or tuff, can sometimes form spectacular layered formations such as these painted hills in central Oregon.

Group: Igneous; Extrusive
Key test(s): Soft, unless welded; angular clasts
Likely locale(s): Downwind from volcanic ranges

Tuff, or lithified volcanic ash, is a common fragmental deposit around volcanoes. The term can be a bit of a catch-all for the pyroclastic material ejected and deposited downwind. Ash fall in a marine environment can result in mixed ash and limestone material, and when hot ash mixes with water, rain, or glacial melt and then rumbles downhill, it is called a lahar. When volcanic ash accumulates in an eruption cloud or column above a volcano and reaches critical mass, it can collapse and rocket down the mountain's flanks to form an ignimbrite. Tuff is usually tan to dark brown in color but can occur as pink, yellow, green, or even purple. Many of the "painted hills" in the western United States are tuffs and volcanic ash. Varieties of tuff include crystal, containing mineral crystals; lithic, with rock fragments; and welded, in which pumice fragments were hot enough to compress in glassy wisps. Usually found as a breccia, with a fine- to medium-grained matrix, tuff can contain large angular material such as pumice, volcanic bombs, and other debris. Tuff beds can be prime fossil-hunting locales, containing petrified wood, leaves, bones, and other organic debris.

Igneous Rocks: Intrusive

The following table compares the various igneous intrusive rocks.

Matrix of Igneous Intrusive Rocks				
Name	Color	Grain Size	Composition	Features
Granite	Light, often pink	Coarse	Feldspar, quartz, mica, and hornblende	Wide range of color; coarse grain size
Syenite*	Light	Coarse	Mostly feldspar and minor mica	Like granite but contains no quartz
Tonalite*	Light to medium; salt-and-pepper	Coarse	Plagioclase feldspar and quartz plus dark minerals	Limited alkali feldspar
Porphyry*	Any	Mixed	Large grains of feldspar, quartz, olivine, pyroxene	Large grains in a fine-grained matrix
Gabbro	Medium to dark	Coarse	High calcium	No quartz; limited olivine
Diorite	Medium to dark	Coarse	Low calcium	Limited quartz
Peridotite*	Dark; often greenish	Coarse	Olivine present	Dense; 40+% olivine
Pyroxenite*	Dark	Coarse	Pyroxene	Rich in pyroxene
Dunite*	Green	Coarse	Olivine dominant	Dense; 90+% olivine
Kimberlite	Dark	Coarse	Source of diamonds	Found in "pipes"
Pegmatite	Any	Very coarse	Usually granitic	Small, local intrusions

= not covered in this guide

9

Granite

Yosemite's El Capitan is a spectacular example of granite.
YOSEMITE PHOTO COURTESY OF BRIAN NAKATA

Group: Igneous; Intrusive
Key test(s): Often pinkish; exfoliation
Likely locale(s): Mountainous terrain

Granite is one of the most common, and recognizable, igneous intrusive rocks. It is usually coarse grained and is primarily composed of feldspar, hornblende, and quartz. It can be massive or display zoning, depending on how close it came to the exterior of the intrusion. Granite is sometimes a salt-and-pepper rock; some granites have a little more feldspar here, while others have a lot more dark hornblende there. Add some mica, some magnetite, and crystalline quartz, all in varying degrees, and it may seem like no two granites are the same.

Granite may occur as minor intrusions, as larger plutons, or as huge batholiths taking up tens of thousands of square miles. It can also form impressive cliffs, such as Yosemite's El Capitan, Mount Rushmore, or the New Hampshire mountains. One field clue is the tendency of granite to weather by exfoliation, where thin sheets peel off like the skin of an onion.

Probably the most interesting aspects of granite are its association with economic ore deposits, often found along the margins of granite intrusions, and the presence of pegmatites.

Gabbro

Gabbro is common in layered intrusives. This specimen is from western Canada.

PHOTO COURTESY OF RICE NORTHWEST MUSEUM OF ROCKS AND MINERALS

Group: Igneous; Intrusive
Key test(s): Dark with coarse grains
Likely locale(s): Mountainous terrain

Gabbro is an interesting intrusive rock made up of feldspar, hornblende, biotite, and magnetite and distinguished by a very coarse texture and a range of dark colors. One helpful identifying characteristic is the presence of green olivine, which contrasts with the usual black or dark gray color. Gabbros can be dark red, however, so that color test isn't precise. Gabbro sometimes makes up the base unit of massive, layered intrusions, such as the famed Stillwater complex in Montana or the Skaergaard intrusion of east Greenland. Perhaps the most famous layered intrusive with gabbro at the base is the storied Bushveld Complex of South Africa, source of much of the world's platinum. These complex intrusions are often rich in chrome as well. Because of the slower crystallization at great depth, olivine crystals tend to sink and form greenish base layers while the rest of the intrusion is still liquid. Layering is sometimes visible in fresh exposures, but gabbro tends to weather faster, and that can also aid in field identification. Other names for gabbro include diabase, dolerite, and black granite. Gabbro takes a polish and can be fashioned into countertops, monuments, and statues.

Diorite

Diorite is a common intrusive igneous rock. This specimen is from Skagway, Alaska.

Group: Igneous; Intrusive
Key test(s): Dark smears, unlike granite; often in dikes and sills
Likely locale(s): Mountainous terrain

Diorite is a medium- to coarse-grained igneous intrusive rock made up of common minerals such as plagioclase feldspar and pyroxene. It is another of the granitic rocks that can be so confusing for beginning rockhounds. It is usually medium-grained, so crystals are recognizable under a hand lens, but it also can be coarse, with zoning common. Diorite is usually gray to dark gray, but it can be lighter, and a green or brown tint isn't out of the question. Salt-and-pepper colors are very common. Compared with granite and granodiorite, plain diorite is not common. It is often found as dikes, sills, and stocks at the margins of large granite batholiths. Dark smears are often present, and these smears help set diorite off from granite. Because diorite is relatively hard, it takes a good polish and was used by ancients for inscriptions, such as the Code of Hammurabi, which was carved into black diorite, and the Rosetta Stone, which was carved into a granodiorite. Economic ore deposits are even more likely in diorite than in granite, especially in associated quartz veins.

Kimberlite

This specimen of Sloan kimberlite is from northern Colorado.

PHOTO COURTESY OF W. DAN HAUSEL; SLOAN-KIMBERLITE.BLOGSPOT.COM

Group: Igneous; Intrusive
Key test(s): Rare garnets; very coarse; large crystals; reacts with acid
Likely locale(s): Older continental crust

Kimberlite is extremely rare, but it's worth noting because of its economic value. Kimberlite is one of the few rocks that contains diamonds (lamprolite is another), and that makes it interesting for rockhounds and prospectors as well as field geologists. While South Africa is the most well-known source of the world's diamonds, there are interesting deposits in North America, including Arkansas, Wyoming, Colorado, Montana, Michigan, and northern Canada. The rock matrix is usually dark gray or dark blue, but it weathers yellow. It is very coarse, with large, rounded crystals easily visible. There is enough calcium carbonate present to bubble when encountering diluted acid. Because kimberlite extends at depth, it is referred to as a "pipe" and is found primarily in cratons—a geological term for stable, continental rock. Kimberlite tends to weather faster than its surrounding host, and geologists have learned to look for round depressions and small ponds or lakes in known kimberlite areas. Geologists can screen alluvial soils and look for kimberlite indicator minerals, such as pyrope garnets, rather than search for actual diamonds.

Pegmatite

This pegmatite specimen, consisting of feldspar and schorl, was found in the Black Hills of South Dakota.
PHOTO COURTESY OF RICE NORTHWEST MUSEUM OF ROCKS AND MINERALS

Group: Igneous; Intrusive
Key test(s): Rare garnets; very coarse crystals
Likely locale(s): Older continental crust

Pegmatites sometimes host the most prized of gems, such as rubies, emeralds, and sapphires, so they are important to learn because of their potential reward. Their name refers to their coarse texture with large crystals, sometimes resembling a patchwork quilt. Pegmatites are usually found at the margins of large granite bodies and contain the same minerals, such as alkali feldspar and quartz. Other minerals found in pegmatite include quartz, tourmaline, topaz, mica, apatite, lepidolite, and monazite. Granite pegmatite and nepheline syenite pegmatite are all usually white because of the amount of feldspar present. Pegmatites can also appear light yellow, tan, or even gray. Zoning is common, with vugs and cavities highly sought for large crystals of quartz, smoky quartz, topaz, and other minerals. The slow cooling process allows large crystals to form, and large cavities provide ample room to develop into larger, perfectly terminated specimens. Famous pegmatite belts in North America include the Black Hills; the Appalachians; Llano County, Texas; and northern Idaho.

Metamorphic Rocks

Use the matrix below as a starting point for understanding how metamorphic rocks are organized.

Name	Hardness	Foliation	Grain Size	Color	Other Details
Soapstone*	Very soft	Foliated	Fine	Light	Greasy
Slate	Soft	Foliated	Fine	Dark	Striking sound
Phyllite*	Soft	Foliated	Fine	Dark	Shiny and crinkly
Serpentinite	Soft	Nonfoliated	Fine	Green	Shiny, mottled
Marble	Medium	Nonfoliated	Coarse	Light	Calcite or dolomite by acid test
Argillite	Hard	Relicts	Fine	Mixed	Common
Greenstone	Medium	Relicts	Fine	Light	Common
Mylonite*	Hard	Foliated	Coarse	Mixed	Crushed and deformed
Schist	Hard	Foliated	Coarse	Mixed	Large, deformed crystals
Gneiss	Hard	Foliated	Coarse	Mixed	Banded
Migmatite*	Hard	Foliated	Coarse	Mixed	Melted
Amphibolite*	Hard	Foliated	Coarse	Dark	Hornblende
Hornfels	Hard	Nonfoliated	Fine or coarse	Dark	Dull, opaque
Eclogite*	Hard	Nonfoliated	Coarse	Red and green	Dense; garnet and pyroxene
Quartzite	Hard	Nonfoliated	Coarse	Light	Quartz—no fizzing

* = not covered in this guide

Slate

Slate sometimes displays small crystals of pyrite.

PHOTO COURTESY OF RICE NORTHWEST MUSEUM OF ROCKS AND MINERALS

Group: Metamorphic
Key test(s): Foliation; visible crystals; density
Likely locale(s): Metamorphic terrain

Increasing metamorphism ⟶

Mudstone	Shale	Slate	Phyllite	Schist	Gneiss

Slate started life as a sedimentary rock, most likely a mudstone, but it has been subjected to heat and pressure that caused the minerals to reorient. At this level of metamorphism, the principal minerals are micas, such as muscovite and chlorite. The presence of garnets is a good indicator that material has metamorphosed further, into a schist. Pyrite crystals and cubes, sometimes weathered to limonite, are common in some slates. Slate is sometimes confused with phyllite, but if you have visible mica flakes and the rock is crumbly and easy to break, it's a phyllite. Slate is usually dark gray, thanks to the presence of graphite from carbonaceous parent material. It can be lighter gray, even green if chlorite is present. The presence of iron can stain slate red, yellow, even brown and purple. Areas where slate is mined for industrial purposes include Vermont, Pennsylvania, and New York in the northeastern United States, parts of Michigan around Lake Superior, plus Georgia and the Rocky Mountains.

Serpentinite

Waxy appearance is common to serpentinite.

Group: Metamorphic
Key test(s): Green; greasy, streaky, and soft
Likely locale(s): Metamorphic terrain

Serpentinite is a collective term for the twenty different magnesium iron phyllosilicates commonly found in varying percentages in most deposits. Serpentine is the most well-known constituent; other common minerals include antigorite, lizardite, and chrysotile. Asbestos is a common constituent, so use caution. This rock is usually green, with streaks of black, yellow, white, or gray. It is very dense, which distinguishes it from soapstone. The best field test is serpentinite's slick, greasy feel, which is a key characteristic. The structure is usually massive, if jumbled, and serpentinite is famous for "slickensides" where microfaulting and zones of movement result in polished, "slick" surfaces. The hardness for serpentinite is usually 4 or less on the Mohs scale, sometimes as low as 2.5, depending on how much hard quartz is present. Plain serpentine is usually softer. Serpentinite is usually found in small pods, lenses, and layers, with steep tilting common. Another interesting field observation is that soils derived from serpentinite usually contain no lime or alkalies, thus no natural fertilizer, and host little vegetation. Quebec, New England, and California have significant serpentinite zones, but many states host exposures.

Marble

This banded marble is from near Ely, Nevada.

PHOTO COURTESY OF RICE NORTHWEST MUSEUM OF ROCKS AND MINERALS

Group: Metamorphic
Key test(s): Hardness; color
Likely locale(s): Metamorphic terrain

There are dozens of varieties of marble, but in general it is simply a limestone or dolomite that has been highly metamorphosed, usually via contact metamorphism. The heat of the metamorphism turns the parent rock into a fine- to medium-grained specimen with a soft, vitreous luster. Marble is normally white— sometimes brilliantly so—but it is often darker thanks to black, green, red, pink, or even yellow staining from various impurities. It often appears shot through with veins of darker material that originated in the limestone bed as clay, sand, silt, or chert. This produces a lined or "marbled" look. Because limestone deposits are often quite large, it's logical that marble is usually massive, with foliation, banding, and streaking all common. Calcite marble effervesces with cold hydrochloric acid. Dolomitic marble must be pulverized and the acid heated to witness the acid effect. Marble is found throughout North America; outcrops are usually found near intrusive bodies or interbedded with mica schists and gneiss. Pure white marble is highly prized for carving and fashioning into statues, such as Michelangelo's famed *David,* created from Carrera marble.

Argillite

This argillite is from the Belt Supergroup, Flathead County, Montana.

PHOTO COURTESY OF RICE NORTHWEST MUSEUM OF ROCKS AND MINERALS

Group: Metamorphic
Key test(s): Foliation; visible crystals; density
Likely locale(s): Metamorphic terrain

Argillite is a common term for any fine-grained sedimentary rocks such as siltstone that have begun to metamorphose. Think of argillites as mud or clay that has started to harden into a rock. Because of the varied nature of their parent rock, argillites can be highly variable in composition but generally have high concentrations of aluminum and silica. Argillites are sometimes mixed with shale layers, making the distinction difficult. Argillite tends to be massive, with bands and zones of different composition common. These rocks are rarely collectible, but there are exceptions. In Idaho's famed Silver Valley, argillites from the Belt Supergroup, mostly in the St. Regis and Revette Formations, host one of the world's major silver deposits, with more than a billion ounces of silver recovered to date. Argillite itself is rarely of such quality that it can be fashioned into carvings or jewelry as can actual catlinite (pipestone). One exception is the hard, fine black Haida argillite of British Columbia, which is still carved by native craftsmen.

Greenstone

This specimen of banded greenstone is from the Pueblo Mountains of northern Nevada and southern Oregon.

Group: Metamorphic
Key test(s): Green
Likely locale(s): Metamorphic terrain

Greenstone is a dense, fine-grained rock that probably started life as a basalt. One nickname is "metabasalt." Greenstone has undergone regional metamorphism. It is generally massive, with occasional evidence of its basalt parentage, such as vesicles. It has a dull luster and is green, pale green, and occasionally yellow-green. The minerals present in greenstone are almost uniformly green—chlorite, epidote, and actinolite predominate. Greenstones also can contain feldspars such as albite. The hardness for greenstone is about 5 to 6 but can vary depending on what minerals dominate the specimen. Outcrops are frequently fractured, resulting in angular, broken debris. Prospectors throughout western North America quickly learned to locate greenstone outcrops, as the telltale green color was easy to spot. Its presence signifies regional metamorphism, and finding greenstone often leads to discovery of quartz and calcite veins in economic ore deposits. However, greenstone can be very common in some areas and is not a sure indicator of success.

Schist

Garnet schist like this specimen can be found in the Wrangell Mountains, Alaska.

PHOTO COURTESY OF RICE NORTHWEST MUSEUM OF ROCKS AND MINERALS

Group: Metamorphic
Key test(s): Foliated mica; garnets; density
Likely locale(s): Metamorphic terrain

Increasing metamorphism ⟶

Mudstone	Shale	Slate	Phyllite	Schist	Gneiss

Schist is one of the more common metamorphic rocks in North America and is relatively simple to identify. Outcrops often contain striking folds and tilted attitudes, along with considerable folding. The thin, platy appearance is classic, and there isn't a good boundary between phyllite and schist. Schist is usually medium to coarse-grained, with mica being obvious. Frequently the mica lines up in a specific orientation, over and over. Gray and brown are common colors, but schist can run to white and yellow, depending on how much iron is present to stain the rock. Schist is quite predictable in its chemical composition and can be used to detect how much metamorphism has occurred. Chlorite and albite schists have undergone low-grade heat and pressure, while the presence of garnet and epidote represent medium-grade metamorphism. Further heat and pressure results in kyanite, a blue, bladed mineral, and staurolite, noted for its twinning. This is the highest grade of metamorphism for a schist. Kyanite is collectible, as is gem-quality garnet.

Gneiss

This banded gneiss is from Georgetown, Colorado.

PHOTO COURTESY OF RICE NORTHWEST MUSEUM OF ROCKS AND MINERALS

Group: Metamorphic
Key test(s): Hard; usually banded
Likely locale(s): Metamorphic terrain

Increasing metamorphism ———→

Mudstone	Shale	Slate	Phyllite	Schist	Gneiss

Gneiss (pronounced *nice*) is another common metamorphic rock. In metamorphic terrain, it can make up most of the pebbles and cobbles of a river or creek. Gneiss is medium to coarse grained, and the minerals are tightly aligned. This rock often shows off light and dark banding in an infinite variety, but it is usually harder than schist and thus sometimes easily distinguished from those rocks. Note, however, that there is no acknowledged cutoff between schist and gneiss. Banding and layering are key characteristics; rocks and outcrops demonstrate both large- and small-scale folding. Look for relict phenocrysts that can get rolled into eye-shaped features referred to by the German name for eye, *augen* gneiss. Accessories include almandine garnet, corundum, and staurolite, but those are more common in schists and will be more like relicts in a gneiss. Look for intense folding, light-and-dark banding, and hard, erosion-resistant rocks at the heart of mountain ranges. Gneiss is common in North American metamorphic terrains.

Hornfels

Spectacular lime hornfels from northern Nevada; it's usually fine grained.

Group: Metamorphic
Key test(s): Foliation; visible crystals; density
Likely locale(s): Metamorphic terrain contacts

Hornfels is a fairly common metamorphic rock in some terrains, but it can be difficult to identify. The name refers to the erosion-resistant Matterhorn in the Swiss Alps, where the rock forms prominent peaks. Biotite hornfels is usually black, or at best dark gray, while lime hornfels is lighter in color. Both varieties are very dense. Hornfels can be very fine grained and massive, with very few signs of relict bedding, as expected in a highly metamorphosed rock. If the sample does show banding, a hornfels will break across those bands, while schist and gneiss still might follow the original foliation. Hornfels is sometimes found in small, round masses just starting to crystallize from heat and pressure, but other deposits can be massive. Hardness ranges from 6 to 7, similar to gneiss. When pounded with a hammer, hornfels breaks into angular splinters, but that isn't a surefire field test. The best places to look for hornfels are around the margins of big batholiths and intrusions, where there was plenty of heat.

Quartzite

This common quartzite pebble shows characteristic fracture across relict bands.

Group: Metamorphic
Key test(s): Hardness
Likely locale(s): Metamorphic terrain

Quartzite is a common pebble in many streams and rivers because it is hard and strong. It started life as a quartz-rich sedimentary rock, but upon metamorphism the clasts reformed as interlocking quartz crystals, resulting in strength against crushing. Quartzite is characterized by a vitreous luster and is usually found as a round, white rock, especially if pure. Other forms of quartzite can be light gray, and it can even run to pink or light brown. Quartzite is usually fine grained, but sometimes crystals are actually visible and the rock can be medium grained. Quartzite deposits are typically thick, massive units, resistant to erosion and forming prominent bluffs. While traces of the original bedding are sometimes obvious—especially if biotite is present, as it tends to line up— quartzite will break across bedding planes, not along them. The hardness, at 7, is typical of quartz, but it can be softer if significant amounts of feldspar and calcite are present. Some collectors confuse quartzite pebbles for agate because they share similarities, but reserve the term *agate* for translucent, banded quartz. The most famous locale is at Quartzsite, Arizona, home to the popular annual rock and gem collectors gathering.

Sedimentary Rocks

Use the matrix below as a starting point for understanding how sedimentary rocks are organized.

Name	Hardness	Grain size	Composition	Features
Coal	Soft	Fine	Carbon	Black; burns
Shale	Soft	Fine	Clay minerals	Splits in layers
Limestone	Soft	Fine	Calcite	Fizzes
Dolomite	Soft	Coarse or fine	Dolomite	Fizzes if powdered
Coquina*	Soft	Coarse	Fossil shells	Bits and pieces
Wacke/ Graywacke	Hard or soft	Mixed	Mixed sediments with rock grains and clay	Gray or dark and dirty
Concretion	Hard	Fine	Cemented grains	Small to large, rounded
Conglomerate	Hard or soft	Mixed	Mixed rock and sediment	Rounded rocks cemented together
Breccia	Hard or soft	Mixed	Mixed rock and sediment	Sharp-edged conglomerate
Mudstone	Hard or soft	Fine	Sand and clay	Very fine
Siltstone	Varies	Fine	Very fine sand; no clay	Gritty
Sandstone	Varies	Coarse	Clean quartz	Grainy
Arkose*	Hard	Coarse	Quartz and feldspar	Quartz sandstone
Chert	Hard	Fine	Chalcedony	No acid fizz

= not covered in detail in this guide

25

Coal

High-grade coal is shiny and brittle. This specimen is from Coal Pass, near Mount Baker, Washington.

Group: Sedimentary
Key test(s): Oily smell; dusty; light heft
Likely locale(s): Metamorphic terrain

Increasing metamorphism ⟶

Peat	Lignite	Bituminous	Anthracite	Graphite	Carbon

Coal actually spans the boundary between sedimentary and metamorphic rocks. Low-grade peat and lignite are basically bog material and thus have a sedimentary origin. But once heat and pressure start to transform those carbon-rich locales, coal gets cooked into a more useful energy source. Even at high grades, coal is very soft; its overall hardness ranges from 1 to 2.5. Once coal bakes into the anthracite stage, it is less greasy, but coal dust is always a problem due to coal's tendency to be brittle and friable. Further metamorphism leads to graphite and pure carbon, but in nature diamonds are created in kimberlite, not from metamorphosed coal. Low-quality coal shows signs of leaves, twigs, and other fossils and displays obvious evidence of original bedding. The higher the quality, the fewer relicts are present, and color can get more interesting, with purple iridescence. Coal is found in seams and zones among sedimentary rocks such as shale and is common in Pennsylvania's coal region and south through the Appalachians.

Shale

This shale is from Wyoming's Bighorn Formation.

Group: Sedimentary
Key test(s): Foliation; visible crystals; density
Likely locale(s): Sedimentary basins

Increasing metamorphism ⟶

Mudstone	Shale	Slate	Phyllite	Schist	Gneiss

Shale is still classified as a sedimentary rock, but it has started to harden in distinguishable ways. Varieties include oil shale, calcareous shale, carbonaceous shale, and more. Shale is usually dark but not always, as its color depends on the source rock it hardened from. Because it has not metamorphosed but simply compacted, it usually shows very thin relict beds, which distinguishes it from the more massive mudstone. Typical grain size for shale is in the silt and clay range. Shale is noted for the way it fractures into plates along bedding planes, where evidence of past water action such as waves, ripples, cracks, and footprints are all observable. Shale beds are easy to spot, although they can easily be confused with platy andesite. Look for evidence of fossils or water features for one differentiator. The most famous fossiliferous shale is the Burgess Shale Formation in British Columbia—a World Heritage site noted for strange, soft-tissue fossils of various evolutionary dead ends from the Middle Cambrian.

Limestone

This fossiliferous limestone is from Wyoming.

Group: Sedimentary
Key test(s): Fizz test; fossils
Likely locale(s): Old sea basins

Limestone is one of the most common sedimentary rocks and comes in many varieties. Strictly speaking, limestone must be at least 50 percent calcium carbonate, as either limestone or aragonite. Impurities are common, such as rock fragments, dolomite, quartz, and clay. Some of the more notable varieties include chalk, coquina, travertine, and oolitic limestone. Chalk is fine grained, derived from the skeletons of tiny sea creatures, while coquina contains large, abundant shell fragments. Travertine is typically banded and colorful. Oolitic limestone refers to the tiny orbs, or oolites, which are basically very small concretions. The term *marl* is used to describe a limestone with a high percentage of silicates. Because it is easily soluble in water, limestone erodes quickly and is marked in the field by pitted and pockmarked outcrops. The hardness for limestone is in the range of 3 to 4. The most reliable test is fizzing in cold, dilute hydrochloric acid. It is susceptible to dissolving in groundwater and sometimes features long underground caves in massive limestone deposits. Two of the more famous limestone cave systems in North America are Carlsbad Caverns in New Mexico and Mammoth Caves in Kentucky, and there are numerous others.

Dolomite

This sample of dolomite is from the Bighorn Formation of Wyoming.

Group: Sedimentary
Key test(s): Fizz test
Likely locale(s): Common

Dolomite, sometimes called dolostone, is typically tan or light gray, but varieties can range to pink and dark gray. It is typically dense, like its cousin limestone, but there is usually little evidence of grains or fossils. Typically dolomite has a very fine texture because it doesn't have the shell fragments or oolites of limestone. Limestone and dolomite share many characteristics, but the big difference is that dolomite has substituted more magnesium for calcium. The science behind creation of dolomite beds is mixed. Geologists have discovered current conditions in tropical environments where dolomite is forming, but other deposits defy easy explanation. With less calcium than limestone, dolomite is slightly more resistant to erosion, but that's not an easy field test. It can appear as bands and grade slowly into limestone. The best test for separating dolomite from limestone is with hydrochloric acid: Dolomite will fizz only if the acid is hot and the material has been ground into a powder, while limestone is much more reactive. Dolomite is used for fertilizers high in calcium and magnesium and as an easy source of magnesium.

Graywacke

This specimen of graywacke is from northern Nevada.

Group: Sedimentary
Key test(s): Clay oxidation
Likely locale(s): Sedimentary basins

Graywacke is an uncommon sedimentary rock composed of rounded or broken bits of shale, slate, basalt, granite, or chert in a fine clay matrix. It is usually gray, hence the name, but can be dark gray or reddish with the presence of iron. There is typically no sorting in a graywacke, giving it an unusual appearance among conglomerates and sandstones. Quartz and feldspar grains are common, but the grain size and structure of graywacke varies quite a bit. Think of it as dirty sandstone, which typically dates to the Silurian period in Europe. Exposures in North America are much less common.

Concretion

This fossil crab (pulalius vulgaris) *was found in Washington.*
PHOTO COURTESY OF RICE NORTHWEST MUSEUM OF ROCKS AND MINERALS

Group: Sedimentary
Key test(s): Round; heavy; splittable
Likely locale(s): Sandstone

Concretions are the lottery ticket of the fossil world. They are typi-cally round or oval structures that can range from tiny, pea-size orbs to beach ball–size or larger monsters. A concretion typically forms when some organic debris starts rolling around in a lime-rich mud within an active marine bay or lagoon, but other chemi-cals that can form concretions include silica and iron. The mud will stick to organic material, and soon the original is completely coated. Further agitation results in a large, round object, which often winds up in sandstone or mudstone, sometimes in striking zones. Not every concretion contains a prize, but many concre-tions do. Fossil hunters value them because the chemical bond holding the concretion together protects the fossil inside, if there is one; thus the snail, crab, or even whale skull is typically well pre-served. Search for concretions in known marine sandstones, silt-stones, and mudstones—especially where the term *cannonball* is used to name local geography.

Conglomerate

The elements of common conglomerate appear glued together.

Group: Sedimentary
Key test(s): Rounded pebbles
Likely locale(s): Nonmarine

Conglomerates are typically found as hard, haphazardly sorted beds of large, rounded pebbles glued together in a fine- to medium-grained matrix. If the pebbles are still angular and edgy, the rock is called a breccia. This is a clastic rock, and the clasts themselves can be big or small. By definition, the rocks and pebbles are greater than 0.08 inch in diameter, but they can range to large cobbles and boulders. Any signs of relict bedding are usually long gone, but sandstones nearby typically show the original orientation of the beds. Because conglomerates are typically deposited in nonmarine waters that are very active, the clasts are usually hard material such as quartzite, chert, or flint, which can survive that much churning. Other plentiful material might include pebbles of basalt, granite, gneiss, and schist. The matrix of a conglomerate is often dominated by silica or calcite, making this rock extremely tough and resistant to erosion. Conglomerates can be worked and reworked repeatedly over time. Fossils are rare in conglomerates because the high velocity required to move these rocks can destroy anything soft.

Breccia

This breccia specimen is from central Oregon.

PHOTO COURTESY OF OREGON DEPARTMENT OF GEOLOGY AND MINERAL INDUSTRIES

Group: Sedimentary
Key test(s): Angular pebbles
Likely locale(s): Nonmarine

Unlike conglomerates, breccias consist of broken rock pieces that are still angular and sharp edged. This is usually evidence that the source material is located nearby, as significant rounding hasn't started. These rocks are typically related to tuff and volcanic ash and are usually lighter in color. They can be made up of mixed material, but the general rule is that the rock fragments are coarse, angular, and varied. There may be little evidence of sorting, with a wide range of large, medium, and small material jumbled together. Breccias are usually found in thick, massive beds, possibly interbedded with fine-grained tuff. Breccias are rarely cemented together with calcite; the usual cementing agent that glues the material together is silica. Breccias are usually seen as evidence of explosive forces, such as meteorite-impact breccias, and of volcanic explosions. Breccias also can depict evidence of significant movement along faults.

Mudstone

This outcrop of fossiliferous Nye mudstone is from the Oregon coast.

Group: Sedimentary
Key test(s): Very fine grained
Likely locale(s): Seashore

Mudstone is a common sedimentary rock that's usually light tan or light gray but can be darker or redder. It is usually dense and sometimes has organic material present in its extremely fine-grained matrix. Mudstone is composed of tiny particles of fine silt or clay, hence the name. Look for clay, such as kaolinite, feldspars, and quartz grains, plus mica flakes, but at a very small scale. When dry, mudstone can appear massive, but when wet the bedding planes are easier to pick out. Mudstones are prone to weather quickly, and their cliffs recede markedly in wet, marine climates. Fossils tend to be well preserved in mudstone, but unless fossils are immediately prepped in the field with preservative, moist air can quickly reduce them to chalk.

Siltstone

The Chuckanut Formation in northwest Washington is the source of this fossiliferous siltstone.

Group: Sedimentary; clastic
Key test(s): Clast size
Likely locale(s): Nonmarine

Siltstone is a common sedimentary rock that generally ranges from light gray to light brown in color but can be darker if more organic material or iron staining is present. It is usually dense and under a microscope reveals smaller particles of silt or clay. Bedding planes are often easy to see, especially if wet. Look for clay, such as kaolinite, feldspars, and quartz grains, plus mica flakes, but at a very small scale. Like mudstones, siltstones weather readily, and their cliffs recede quickly in wet climates. Fossils tend to be poorly preserved in these rocks and often must be immediately prepped in the field with Vinac, a polyvinyl acetate coating.

Sandstone

The cliff dwellings at Mesa Verde, Colorado, are tucked into dramatic sandstone cliffs.

Group: Sedimentary; clastic
Key test(s): Sandy feel
Likely locale(s): Nonmarine

Sandstone is one of the most common sedimentary rocks and is found throughout North America. It comes in many varieties, such as arkose, which is rich in silica. Frequently sandstones in the field are found interbedded with mudstone, limestone, shale, and other sedimentary rocks. By definition, all sandstones are marked by a grain size of 0.05 to 2.0 mm, which makes it medium grained compared to mudstone and siltstone but not coarse like a conglomerate. Sandstone is quite varied in color and is found in hues of gray, brown, red, yellow, and even white. Sandstone bluffs and beds usually show at least some evidence of sorting, making grains uniform in size throughout a zone. Bedding is common. Sandstone is usually deposited horizontally but can show angular beds on a microscale. Quartz is the most common component due to its hardness, but feldspar is usually present in a high percentage. Silica usually holds the clasts or grains together in a sandstone, but calcite or even iron oxide will also serve. Sandstone is relatively easy to shape and fashion into buildings and walls, and it is often used as a decorative stone. Fossils are common in sandstones, such as the dinosaur fossils of the Morrison Formation in the Rocky Mountains.

Chert

This specimen of red chert was found in northern Nevada.

Group: Sedimentary
Key test(s): Hardness 7; organic nature; ooze origins
Likely locale(s): Between basalt flows

Like most quartz incarnations, chert has a hardness of 7. It is commonly white in color, but banded chert can take on reds, yellows, and even greens. It can easily be confused with jasper for that reason, but chert is typically a by-product of organic ooze from the seafloor hardened into a rock, while jasper results from circulating silica solutions, usually in basalt. Some jasper derived from silica-rich ash flows is also called chert, so there is some controversy. At high magnification some cherts display tiny little skeletons. Chert also forms when silica-rich solutions replace limestone, further confusing things. When black, chert is more commonly referred to as flint and was easily fashioned into tools and weapons by Stone Age artisans. Chert is dense; smooth when polished, it can feel very rough where exposed. It can occur as massive cliffs of considerable thickness but is more common as nodules, lenses, and interbedded zones. Bedded cherts in basalts and sandstones are prized by collectors, who look for seams of material big enough to slice with a rock saw and polish into cabochons.

PART 2:
MINERALS

Common Minerals

Agate

Banded agates, known as "lakers," are highly collectible in the Lake Superior region of Minnesota.

Quartz, SiO_2
Family: Cryptocrystalline quartz
Mohs: 6½–7
Specific gravity: 2.65
Key test(s): Banding; translucence
Likely locale(s): Veins and blebs in basalt

Agate is a label loosely used by rockhounds to describe any clear or translucent quartz, but the name is interchangeable with *chalcedony,* which is the preferred term among mineralogists. Agate is one of the most popular and collectible forms of cryptocrystalline quartz, and it comes in a variety of forms, including moss agate and banded agate. Interestingly, many of the most famous agate, such as Oregon's Holley blue and Washington's Ellensburg blue agate, are technically forms of chalcedony. Generally *agate* should be reserved for translucent, banded chalcedony, and usually the banding is quite distinctive. Typical colors are clear, red, yellow, or a pleasing, highly sought-after light blue. True agate has a fine grain and excellent color. Lake Superior banded agates are some of the most famous in North America and are derived from the top layers of basalt where vesicles are replaced with silica. Spectacular banded fortification agates are also found in Kentucky. Arizona has fire agate.

Apatite

This blue-gray apatite is from Arizona.
PHOTO COURTESY OF JIM SMITH, "IT'S JUST A ROCK" COLLECTION

Calcium phosphate, $Ca_5(PO_4)_3(F,Cl,OH)$
Family: Phosphates
Mohs: 5
Specific gravity: 3.1–3.2
Key test(s): Hardness test
Likely locale(s): Pegmatites

Apatite is the name given to a group of phosphate minerals such as fluorapatite and chlorapatite, so the chemical formula can vary according to how much of one element or another is present. As expected, the color varies. Darker green, purple, or violet are common, as are red, yellow, and pink. Apatite is a common mineral in igneous rocks; larger crystals occur in pegmatites. Fluorite has similar coloration but isn't as hard, and the crystal habit is cubic; quartz is harder. Interestingly, apatite is the hardening agent for bones and teeth. Noted locales include the Himalaya Mine in southern California; Pelham, New Hampshire; and within marble deposits near Ottawa, Canada. Another noted North American apatite locale is at Ciudad Durango in Mexico.

Augite

These loose augite crystals and augite in matrix are from the Oregon Coast Range.

Silicate, Ca,Na(Mg,Fe,Al)(Al,Si)$_2$O$_6$
Family: Silicates
Mohs: 5–6½
Specific gravity: 3.2–3.6
Key test(s): Black, stubby crystals
Likely locale(s): Volcanic dikes

Augite is a common mineral in the pyroxene group and is related to wollastonite, diopside, hedenbergite, and pigeonite. Augite typically forms short, stubby crystals in the monoclinic system, but they are usually too small to detect without a hand lens. Large crystals do form occasionally and can be striking and quite shiny, but they pit quickly when exposed to air. Augite has a greenish-white streak, which is a good field test. Augite is commonly found in igneous rocks, sometimes as an essential mineral in dikes. One of the better locales in North America is at Cedar Butte in northwest Oregon.

Calcite

This calcite crystal was found in Yamhill County, Oregon.

PHOTO COURTESY OF RICE NORTHWEST MUSEUM OF ROCKS AND MINERALS

Calcium carbonate, CaCO$_3$
Family: Carbonates
Mohs: 3
Specific gravity: 2.7
Key test(s): Rhombohedral angles; hardness
Likely locale(s): Veins

Calcite is very common and is found as crystals in pegmatites, as veins in basalts, and mixed with zeolites. It is usually yellow but can occur in a pure, clear crystal form. It has a white streak, and its luster is vitreous but sometimes very dull. One of the most classic field tests for calcite is the rhombohedral crystal habit; calcite will cleave easily along those planes. Calcite also fluoresces. Gypsum doesn't effervesce in acid, and quartz is much harder than calcite. Calcite sometimes forms massive crystals, but it can also be fibrous. Two well-known crystal forms are dog-tooth and nail-head calcite. Perfect calcite crystals display double refraction—in essence doubling the image they are set upon. Calcite is common to igneous, metamorphic, and sedimentary rocks and is common throughout North America.

Chalcedony

This sample of lavender-blue chalcedony is from Holley, Oregon.
PHOTO COURTESY OF JIM SMITH, "IT'S JUST A ROCK" COLLECTION

Quartz, SiO$_2$
Family: Cryptocrystalline quartz
Mohs: 7
Specific gravity: 2.6–2.64
Key test(s): No banding
Likely locale(s): Quartz-rich areas

Chalcedony is one of the most common forms of quartz and is an all-inclusive term for a vast variety of collectible material, including agate, jasper, sard, and carnelian. When red or orange, chalcedony is often referred to as sard, or carnelian. Other varieties include layering, making for sardonyx; green with red spots is called bloodstone, or heliotrope. When the material is banded or clear, rockhounds use the term *agate;* if there are inclusions the sample is called moss agate. Apple-green chalcedony is called chrysoprase. Red, yellow, or brown varieties are called jasper, and when the material is white or gray, we call it flint. There is no cleavage, as there is no crystal habit, and the fracture is conchoidal. Chalcedony typically has a waxy, vitreous, or dull luster. Chalcedony is found in virtually all alluvial gravels and beach deposits to some degree.

Carnelian

This banded red-orange carnelian with drusy quartz interior is from Lucas Creek, Washington.

Quartz, SiO$_2$
Mohs: 7
Specific gravity: 2.6
Key test(s): Color
Likely locale(s): Seams within basalt flows

Carnelian is another cryptocrystalline form of quartz, with small amounts of iron giving it a brownish-red tint. Basically it is a special form of chalcedony. The term *sard* is used to denote carnelian that is darker and harder, but both can be present in the same specimen. Carnelian is as hard as quartz and agate, 7 on the Mohs scale, and it has a white streak. Carnelian also has a characteristic conchoidal fracture pattern. Because of its hardness, carnelian tends to remain in stream and river gravels, and rockhounds often search for it along ocean beaches.

Common Opal

Common opal like this specimen can be found within diatomaceous earth deposits.

Quartz, SiO_2*nH_2O
Family: Cryptocrystalline quartz
Mohs: 5½–6
Specific gravity: 2.0–2.2
Key test(s): Softness; lighter than other quartz
Likely locale(s): Quartz-rich areas

Opal is another form of quartz but this time with considerable water present. There are numerous varieties of common opal, and it is widespread throughout North America. Common opal, sometimes referred to as hyalite, has a distinctive vitreous, pearly luster. It is usually white, but other colors include yellow, brown, green, blue, and gray. Its streak is white. Opal is not crystalline; there is no habit or cleavage, and it has a conchoidal fracture. Quartz is harder and lacks the conchoidal fracture pattern of opal. Jasper and agate are both harder as well. Opal frequently forms in cavities, fractures, and air bubbles in basalt, where it sometimes weathers out and remains as small, round pea-size blebs. Wood and shells often become "opalized" through replacement. Opal will fluoresce, which also sets it apart from jasper and quartz. Common opal is sometimes associated with diatomaceous earth pits, where it can be a nuisance for miners.

Crystalline Quartz

Here are three varieties of crystalline quartz, including dark smoky quartz, purple amethyst, and clear quartz crystals.

Quartz, SiO_2
Mohs: 7
Specific gravity: 2.65
Key test(s): Color
Likely locale(s): Seams within basalt flows; pegmatites

Quartz crystals are distinctly hexagonal, with pyramidal termina-tions. Quartz crystals are most prized when displaying water-clear color and perfect pointed ends, but other varieties, such as ame-thyst (purple) and smoky quartz (medium brown to dark black) are popular. Like most quartz, the streak for crystalline quartz is white. There is no cleavage in quartz crystals, which have a vit-reous, greasy luster. Crystalline quartz is softer than corundum, topaz, and diamond but harder than calcite. Gem-quality quartz crystals are associated with granite pegmatites, but common quartz is abundant and found in all three classes of rock. The "Herkimer diamonds" of New York are exquisitely terminated pure crystalline quartz—*not* diamonds.

Epidote

This "jackstraw" epidote is from Prince of Wales Island, Alaska.

PHOTO COURTESY OF RICE NORTHWEST MUSEUM OF ROCKS AND MINERALS

$Ca_2Al_2FeOSiO_4Si_2O_7(OH)$
Family: Silicates
Mohs: 6–7
Specific gravity: 3.3–3.6
Key test(s): Color
Likely locale(s): Contact metamorphism

Epidote is a common rock-forming mineral with a characteristic green color, although it can occur as yellow-green or even greenish-black. Epidote has a colorless to gray streak. Crystals are common in the monoclinic system, forming long, slender crystals or as massively tabular encrustations. Actinolite is also green but cleaves in two directions. Olivine is slightly heavier. Epidote occurs in many situations: in epidote schist, in granite pegmatites, in basalt cavities, along with andesite, in greenstones, and in contact metamorphic rocks. Alaska and Connecticut host significant epidote occurrences.

Fluorite

These green fluorite crystals on a drusy quartz background are from the Rock Candy Mine, Grand Forks, British Columbia.

PHOTO COURTESY OF BOB JACKSON; GEOLOGYADVENTURES.COM

Calcium fluoride, CaF$_2$
Family: Halide
Mohs: 4
Specific gravity: 3.0–3.6
Key test(s): Pastel purple, pink, green color; cubic crystal habit
Likely locale(s): Found in hydrothermal, sedimentary, igneous, and volcanic deposits

Fluorite comes in a variety of colors such as purple, green, and pink; banding is common. It leaves a white streak, as do many minerals, so a better test is the hardness test or the isometric cubic crystals. Fluorite crystals are rare in that they cleave perfectly in four directions. Fluorite has a vitreous luster and is softer than quartz but harder than calcite. Another good field test is to check for fluorescence. Fluorite veins are often associated with economic ore deposits, such as galena and sphalerite. There are numerous fluorite locales across North America; one of the largest is in Newfoundland, Canada. The most popular collecting site is a fee-dig at the Rock Candy Mine in British Columbia.

Gypsum

These gypsum crystals are from the "Cave of Swords" in Chihuahua, Mexico.

PHOTO COURTESY OF RICE NORTHWEST MUSEUM OF ROCKS AND MINERALS

Calcium sulfate, $CaSo_4*2H_2O$
Mohs: 1½–2
Specific gravity: 2.3
Key test(s): Softer than calcite
Likely locale(s): Hydrothermal basins

Gypsum is usually white in hand specimens, but it sometimes appears colorless in pure crystals. It leaves a white streak. Crystals are often diamond shaped, especially in the form of selenite, and a rhombic pattern is common, as is a massive, fibrous "satin spar" form. Cleavage is perfect in one direction. Gypsum is softer than calcite, which is an easy field test, as both can occur as rhombohedrons. Gypsum is common in hydrothermal replacement deposits and in sedimentary basins, where it is sometimes strip-mined in large quantities for use in sheet rock or as fertilizer. Gypsum sometimes forms "desert roses" as radiating, stubby blades mixed with sand grains and can form "daisy gypsum" in a twinned, radiating spray. One form of gypsum, selenite, was discovered in a Mexican cave in the form of 30-foot transparent crystals. There are numerous gypsum deposits across North America, with Wyoming a major contributor.

Halite

This specimen of cubic halite was found in southern California.

PHOTO COURTESY OF RICE NORTHWEST MUSEUM OF ROCKS AND MINERALS

Sodium chloride, NaCl
Family: Chlorides
Mohs: 2–2½
Specific gravity: 2.2
Key test(s): Taste
Likely locale(s): Evaporite deposits

Halite, or rock salt, is usually clear but is sometimes tinted pink or gray. It leaves a white streak. Halite is strongly isometric, and when pure it features well-formed crystals in tiny cubes. It has a vitreous luster and has perfect cleavage in three directions. Halite is similar to cryolite but harder and has a salty taste. Halite is typically found in large evaporite deposits and in large domelike underground deposits. Many deserts in the Basin and Range province of the western United States feature halite-encrusted shorelines, such as the Great Salt Lake of Utah. These dried-up bodies of water often contain halite or other salts.

Hornblende

These coarse, black hornblende crystals are from Nevada.

$(Ca,Na,K)_{2-3}(Mg,Fe^2+,Fe^3+,Al)_5(SiAl)_8O_{22}(OH)_2$
Family: Inosilicates
Mohs: 5–6
Specific gravity: 2.9
Key test(s): Cleavage angles
Likely locale(s): Common rock-forming mineral

Hornblende is a common rock-forming mineral, usually black but sometimes green or brown. It is actually the name given to a series of amphiboles, all differing by the amounts of iron, magnesium, calcium, sodium, aluminum, and potassium they contain. Hornblende leaves a colorless streak; crystals are monoclinic, usually short, and prismatic. Cleavage is perfect in two directions. Hornblende can be confused with schorl, which is black tourmaline, but schorl doesn't cleave like hornblende. Diamond-shaped cleavage angles and a greenish-black color are the best field indicators. Hornblende is rarely found as collectible crystals.

Jasper

This picture jasper, known as "morrisonite," is from western Idaho. Most jasper is far less dramatic.

Quartz, SiO$_2$
Family: Cryptocrystalline quartz
Mohs: 7
Specific gravity: 2.65
Key test(s): Conchoidal fracture
Likely locale(s): Basalt flows

Jasper is another common form of quartz. Typically derived from silica-rich ground solutions circulating through basalt, it's technically a form of chalcedony. Jasper comes in many shades and colors but is usually red, tan, or yellow and sometimes green. It has no crystal habit and leaves a white streak. Gem-quality jasper is hard and has a conchoidal fracture and was sometimes used in prehistoric times for tools and arrowheads. Poor-quality jasper is often porous on the surface and thus does not take a polish. Common jasper is found in most regions where basalt is common, but not always. Noted locales for picture jasper include Biggs, Oregon; the Owyhees; and Bruneau, Idaho.

Mica Group

This sample of muscovite mica is from Mica Mountain, northern Idaho.

Muscovite, $KAl_3Si_3O_{10}(OH)_2$
Biotite, $K(Mg,Fe)_3(Al,Fe)Si_3O_{10}(OH,F)_2$
Family: Phyllosilicates
Mohs: 2–3
Specific gravity: 2.7–3.0
Key test(s): Platy cleavage
Likely locale(s): Schists

Increasing metamorphism ⟶

Chlorite	Biotite	Garnet	Staurolite	Kyanite	Sillimanite

The mica group is composed of several similar sheetlike minerals, including muscovite, biotite, phlogopite, lepidolite, and chlorite. Of the two most common varieties, muscovite is white or colorless; biotite is much darker. Chlorite is greenish, while lepidolite is pink or light purple. Mica has a vitreous, pearly luster and leaves a colorless streak. Perhaps most characteristic, its cleavage is perfect in one direction, sometimes resulting in great sheets big enough to use as windows for wood stoves. Because mica flakes can look yellow and shiny, they are sometimes mistaken for fool's gold. Micas are found in granitic pegmatites, a historic source of muscovite, while phlogopite occurs in marble and hornfels. Biotite is common in schists, while lepidolite is restricted to granitic pegmatites.

Olivine

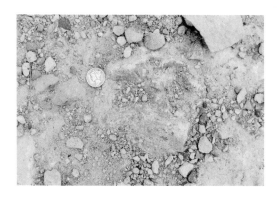

This olivine-rich basalt is from Dunraven Pass, Yellowstone National Park.

$(Mg,Fe)_2SiO_4$
Family: Nesosilicates
Mohs: 6½–7
Specific gravity: 3.3–3.4
Key test(s): Green color
Likely locale(s): Found with metamorphic rocks and in hydrothermal replacement deposits

Olivine is usually dark green, with a vitreous luster and colorless streak. When found in crystalline form, such as the semiprecious gemstone peridot, olivine will cleave in two directions, but it is more commonly associated with olivine-rich basalt, which can range to almost bluish-green. Olivine is the name for a series of minerals, with forsterite on the magnesium-rich end of the series, and by far the most common, and fayalite on the iron-rich end. Olivine is actually a common rock-forming mineral, composing much of the earth's mantle, and there are beaches in Hawaii that have turned green from the presence of so much olivine. In addition, some rare meteorites were discovered to contain olivine crystals, so have moon rocks and Mars samples, plus certain comets.

Petrified Wood

Samples are from Saddle Mountain, Washington. Standing log is at Yellowstone National Park.

SiO$_2$
Family: Silicates
Mohs: 7
Specific gravity: 2.2–2.6
Key test(s): Rings, lines, and wood features
Likely locale(s): Volcanic ash deposits, sedimentary rocks

Petrified wood is the common name for fossil wood, and is used interchangeably with opalized wood, agatized wood, and silicified wood. Whatever the name, the conditions under which wood is turned to stone involve replacement of the original cell structure with quartz. Petrified wood comes in a wide variety of colors and conditions. It is usually black, red, white, or yellow, depending on contamination from iron, carbon, copper, cobalt, and other impurities, and also depending on the plant species it originated from. Some samples are very hard and take an excellent polish, but other samples may be porous and unsuitable for jewelry. At Virgin Valley, Nevada, rockhounds have unearthed opalized wood and bone with fantastic rainbow colors. Other famous locales for the more common petrified wood include Petrified Forest National Park in Arizona, Ginkgo Petrified Forest State Park at Vantage, Washington, and Escalante Petrified Forest in Utah.

Potassium (Alkali) Feldspars

Microcline is a common alkali feldspar. This specimen is from near Helena, Montana.

Orthoclase, $KAlSi_3O_8$
Anorthoclase, $(Na,K)AlSi_3O_8$
Sanidine, $(K,Na)AlSi_3O_8$
Microcline, $KAl AlSi_3O_8$
Albite, $NaAlSi_3O_8$
Family: Tectosilicates
Mohs: 6–6½
Specific gravity: 2.5–2.6
Key test(s): Hardness
Likely locale(s): Common rock-forming mineral

Alkali feldspars, sometimes known as potassium feldspars or potash feldspars, are usually white, light gray, or light pink, but some varieties appear light blue, green, yellow, or red. Microcline is often tinted bluish-green. Potassium feldspars form a series between orthoclase and albite; albite is also considered an end member of the plagioclase feldspars. All feldspars are key rock-forming minerals for igneous and metamorphic rocks and thus quite common, making up more than 50 percent of the earth's crust. There are semiprecious gem varieties: Amazonite is a light blue gem form of the potassium feldspar microcline. Orthoclase feldspar is sometimes referred to as moonstone.

Plagioclase Feldspars

These orthoclase crystals in matrix are also from Montana, from the collection of Bill Porter.

Albite, $NaAlSi_3O_8$
Oligoclase, $(Na,Ca)(Al,Si)AlSi_2O_8$
Andesine, $NaAl AlSi_3O_8$
Labradorite, $(Ca,Na)Al(Al,Si)Si_2O_8$
Bytownite, $(Na,Si)(CaAl)AlSi_2O_8$
Anorthite, $CaAl_2Si_2O_8$
Family: Tectosilicates
Mohs: 6–6½
Specific gravity: 2.6–2.8
Key test(s): Hardness
Likely locale(s): Common rock-forming mineral

Plagioclase feldspars are noted by how much anorthite is present, ranging from 0 percent for albite to 100 percent for pure anorthite. These minerals have a vitreous luster and a white streak. The color is variable but typically dull white or light gray, with light tinting common. Labradorite is often green or blue. Plagioclase feldspars only crystallize in the triclinic system, whereas potassium feldspars crystallize in both the monoclinic and triclinic systems. Adularia is a transparent and opalescent form of orthoclase called moonstone. Sunstones are a form of andesine. Labradorite is another gem variety of feldspar.

Rhodonite

Manganese silicate, $MnSiO_3$

Family: Metal silicates

Mohs: 5½–6½

Specific gravity: 3.6–3.8

Key test(s): Pink (when fresh); tiny "roads"

Likely locale(s): Found with metamorphic rocks and in hydrothermal replacement deposits

Rhodonite is usually easy to identify because it is rose pink in the field and is one of the few minerals found in that color. Rose quartz is harder than rhodonite and usually more translucent. Just to confuse matters, some rhodonite samples are red, brown, or yellow, so that test isn't foolproof. Rhodonite is translucent and leaves a colorless or white streak, and the luster is vitreous. Crystals are not common but appear in the triclinic habit and are typically large and flattened. Most rhodonite samples are massive, with good cleavage at nearly right angles and black manganese oxide fillings that look like roads—a useful memory device. Typical fracture patterns are conchoidal, as with many other minerals. The classic collecting locale for rhodonite is at Franklin Furnace, New Jersey, where mines have given up pink crystals in large, tabular shapes.

Staurolite

This staurolite schist with crystals, including twinned diagonal cross, is from near Avery, Idaho.

$Fe_2Al_9Si_4O_{22}(OH)_2$
Family: Neosilicates
Mohs: 7–7½
Specific gravity: 3.7–3.8
Key test(s): Hardness; twinning
Likely locale(s): Highly metamorphosed schists

Increasing metamorphism ———→

Chlorite	Biotite	Garnet	Staurolite	Kyanite	Sillimanite

Staurolite, which is mainly found in staurolite schist, is an indicator mineral for intense metamorphism. A staurolite schist denotes a regionally metamorphosed rock between garnet and kyanite grade. Staurolite crystals are usually brown, with a vitreous luster and a white streak. The crystal habit is monoclinic; twinning is common, such as right-angle "fairy" crosses. Cleavage is poor, in lengthwise direction. Two noted locales for excellent cross-shaped staurolite twins are in Georgia and Tennessee, with other collecting locales in New England, Pennsylvania, and North Carolina.

Talc

This talc-rich soapstone is from the Lake Wenatchee area, Washington.

$Mg_3Si_4O_{10}(OH)_2$
Family: Magnesium silicates
Mohs: 1
Specific gravity: 2.6–2.8
Key test(s): Soft
Likely locale(s): Metamorphic terrain

Talc is easy to test in the field because you can scratch it with your fingernail. It is the main component of most soapstones and is usually found as a mix of magnesium silicates, such as tremolite or magnesite. Beware of a tendency for asbestos to accompany talc. Color varies depending on associated minerals present and can range from white to bright green. The streak is white. Talc has a greasy feel and shows a pearly luster. Crystals are extremely rare. Look for talc in metamorphic regions with ultramafic rocks such as dunite or peridotite. One prime area is the belt of metamorphic rocks in western California extending north from Santa Barbara.

Tourmaline

This specimen of "watermelon" tourmaline (elbaite) is from the Himalaya Mine in southern California.
PHOTO COURTESY OF JESSICA SCHENK, HIGH DESERT GEMS & MINERALS

$Na(Mg,Fe)_3Al_6(BO_3)_3(Si_6O_{18})(OH,F)_4$
Family: Cyclosilicates
Mohs: 7–7½
Specific gravity: 3.0–3.3
Key test(s): Harder than apatite
Likely locale(s): Pegmatites

Tourmaline comes in a variety of forms. The most common is black schorl, which is pictured on the page for pegmatite. Semiprecious gem varieties include elbaite, first noted on Elba Island but now primarily known from southern California; indicolite, which is blue; rubellite, which can be pink or red, and dravite, which is brown. Tourmaline has a vitreous luster, leaves a white streak, and forms a hexagonal crystal with no cleavage. Hornblende is also black, but schorl has a triangular cross section. Gem-quality tourmalines are usually found in vugs and cavities within granite pegmatites, such as sites in Maine and southern California.

Zeolite Group: Heulandite

This world-class huelandite is from Challis, Idaho.

(Ca,Na)(Al$_2$Si$_7$O$_{18}$)*6H$_2$O
Family: Zeolites
Mohs: 3–3½
Specific gravity: 2.1–2.2
Key test(s): Pink (when fresh)
Likely locale(s): Hydrothermal fillings

Heulandite is the name for a group of zeolites known for a distinctive pink-colored form, but often crystals are white or lack color completely. Crystals are monoclinic and sometimes display a unique coffin shape. Luster can be pearly or vitreous. Heulandite is usually found with other zeolites in veins or pockets within basalt and andesite, but it can also occur in schists. As with all zeolites, the definitive text is *Zeolites of the World* by Rudy Tschernich. One prime collecting area is near Challis, Idaho.

Zeolite Group: Stilbite

Stilbite crystals are usually short, stubby, and milky white.

$NaCa_2Al_3Si_{13}O_{36}*16H_2O$

Family: Zeolites

Mohs: 3½–4

Specific gravity: 2.12–2.22

Key test(s): Hardness

Likely locale(s): Hydrothermal fillings

Like heulandite, stilbite is the name for a series of zeolites rather than a single type; the difference is in the ratio of sodium and calcium present. Most stilbites are richer in calcium. Crystals are usually colorless or white but occasionally can be pink. Iceland has long been a prime source of excellent stilbite specimens, but stilbite is a common zeolite and is found alongside the rest of the zeolite family in vugs and cavities.

MINERALS

Metallic Minerals

Copper

This copper mass was found on the Keweenaw Peninsula, Michigan.

PHOTO COURTESY OF RICE NORTHWEST MUSEUM OF ROCKS AND MINERALS

Copper, Cu
Family: Elemental metal
Mohs: 2½–3
Specific gravity: 8.9
Key test(s): Color; softness
Likely locale(s): Hydrothermal

When fresh, native copper can appear as bright as a shiny new penny. However, copper quickly tarnishes when exposed to air and starts to turn black, green, or blue. Native copper nuggets leave a characteristic copper streak, and this mineral is pretty easy to identify thanks to the copper coins in our financial system. Crystals are rare and in the isometric habit. Copper is noted for its malleability, meaning it bends and can be pounded, flattened, or rolled. The most famous locale for native copper is Michigan's Keweenaw Peninsula. Other areas with at least interesting native copper veins, stringers, and showings include Pima County, Arizona; New Mexico; Oregon; Alaska; New Jersey; and Nova Scotia.

Chalcopyrite

These chalcopyrite crystals are from Cornwall, England.

Copper iron sulfide, $CuFeS_2$
Family: Metal sulfides
Mohs: 3½
Specific gravity: 4.1–4.3
Key test(s): Noncubic crystals
Likely locale(s): Known copper areas

Chalcopyrite is a common iron sulfide that is chemically quite similar to pyrite except that chalcopyrite contains a single copper atom in place of an iron atom; pyrite has no copper. Like pyrite, chalcopyrite is brassy and golden yellow, but chalcopyrite is softer than pyrite, and its crystal habit is a tetrahedron rather than cubic. Its streak is greenish-black, also similar to pyrite. Gold is softer and malleable; therefore chalcopyrite is rarely considered "fool's gold." Chalcopyrite is often associated with hydrothermal economic ore deposits that host silver and gold. It is usually found in massive sulfide zones with other sulfides, especially pyrite, and is a primary copper ore. It occurs in huge masses within the Temagami greenstone belt in Ontario, Canada.

MINERALS

Pyrite

Pyrite such as this cube and mass might be mistaken for gold by novice prospectors.

PHOTO COURTESY OF TIM FISHER; WWW.OREROCKON.COM

Iron sulfide, FeS$_2$
Family: Metal sulfides
Mohs: 6–6½
Specific gravity: 4.9–5.1
Key test(s): Hardness
Likely locale(s): Known sulfide areas

Pyrite, or iron pyrite, is also known as "fool's gold" because of its brassy, yellowish color. Pyrite generally forms in the cubic crystal habit and often has striations, or lines, on its crystal faces. Pyrite crystals sometimes twin, interlock, and form interesting masses. Pyrite tarnishes rapidly, becoming darker and somewhat iridescent as oxygen attacks the iron-sulfur bond. One sure field test is streaking: Pyrite leaves a brownish-black or even greenish-black streak that smells faintly of sulfur, depending on how big a streak you make. Pyrite is hard, registering as high as 6½ on the Mohs scale, just below quartz. Many forms of pyrite are highly collectible, but when collected from a sedimentary environment, this mineral can deteriorate over time—a tendency known as "pyrite disease." Other than cubes, pyrite is also notable for forming pyrite "sand dollars," which aren't fossils but sprays formed between layers of slate.

Malachite

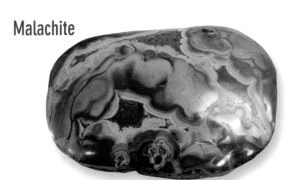

This polished malachite is from Arizona.

PHOTO COURTESY OF RICE NORTHWEST MUSEUM OF ROCKS AND MINERALS

Copper carbonate, $Cu_2CO_3(OH)_2$
Family: Metal carbonates
Mohs: 3½–4
Specific gravity: 3.6–4.0
Key test(s): Green; presence of azurite
Likely locale(s): Copper deposits

This mineral can be a welcome sign in otherwise perplexing or barren outcrops. Even in low concentrations, malachite leaves a telltale green stain, indicating that there is some form of mineralization present. Since it is a carbonate, it is often associated with limestones, which are calcium carbonate. In a pure form, malachite makes for a nice specimen and will take a high polish. The key characteristic for identifying malachite is its bright green appearance; it can occur in shades of green from light to dark, even black-green. Malachite also has a light green streak. Malachite often displays banding, especially when prized malachite stalactites are sliced horizontally. Arizona's copper-mining region is probably the most notable North American locale for malachite, but every mining region known for copper will display at least superficial malachite staining.

Azurite

This sample of velvety azurite is from Greenlee County, Arizona.

PHOTO COURTESY OF RICE NORTHWEST MUSEUM OF ROCKS AND MINERALS

Copper carbonate, $Cu_3(CO_3)_2(OH)_2$
Family: Metal carbonates
Mohs: 3½–4
Specific gravity: 3.7–4.0
Key test(s): Color; hardness
Likely locale(s): Copper deposits

Azurite is a striking blue mineral closely associated with green malachite in copper-mining regions. The two make a pleasing combination when fresh, but azurite tends to lose its brilliance if subjected to heat, bright light, or simply too much air over too long a time. When fresh it has a bright blue streak, which was used for paint pigments during the Middle Ages in Europe. Color and streak are the two key characteristics when identifying field samples. Two well-known locales for azurite are Bisbee, Arizona, and Sonora, Mexico, but most known copper regions will host a lot of malachite and a little azurite.

Bornite

This specimen of bornite, or "peacock" ore, is from Helena, Montana.

Copper iron sulfide, Cu_5FeS_4
Family: Metal sulfides
Mohs: 3–3¼
Specific gravity: 4.9–5.3
Key test(s): Peacock color; heft
Likely locale(s): Rich copper deposits

Bornite is known as the "peacock" ore because it has so many red, purple, bronze, blue, and other hues, all calling for attention at once. Rich in copper by weight, it is an important sulfide of copper. Bornite has a metallic luster, and its streak is grayish black, but you probably won't need to streak it for identification purposes thanks to the striking colors. Bornite crystals are rare; this mineral generally appears as disseminated deposits in skarns, veins, and massive sulfide deposits. Two prominent locales known for excellent bornite crystals are near Butte, Montana, and Bristol, Connecticut. Most sulfide districts in North America with any appreciable copper have some bornite showings; one prominent locale is in the Seven Devils region of western Idaho, above Hells Canyon.

Hematite

This botryoidal hematite was found in the Lake Superior area.

Iron oxide, Fe_2O_3
Family: Metal oxides
Mohs: 5–6
Specific gravity: 4.9–5.3
Key test(s): Bright red streak
Likely locale(s): Iron-rich mineralized areas

Hematite is rich in iron and is the most common iron ore due to dominating banded iron formations. In the field hematite can appear gray or black and has a distinct metallic luster, but after prolonged exposure to air, hematite will eventually start to show its characteristic red signature. Hematite has a striking red streak, which is one telltale identifier. Crystals are varied; they can be hexagonal, tabular, or columnar. Hematite can also display a rounded, bubbly botryoidal habit. Limonite, ilmenite, and magnetite are all similar in appearance, but the deep red streak sets hematite apart from imposters. Hematite can be found as red ochre, one of the oldest color tints known to man. Yellow ochre is also hematite, but it contains extra water that results in a yellow color. The most significant hematite deposit in North America is the banded iron found in Minnesota's giant Mesabi Range. The planet Mars is colored red by an abundance of fine-grained hematite.

Magnetite

This sample of magnetite is from Pennsylvania.

Iron oxide, Fe_3O_4
Family: Metal oxides
Mohs: 5½–6½
Specific gravity: 5.1–5.2
Key test(s): Magnetic; attracts magnetic dust
Likely locale(s): Iron-rich mineralized areas

Magnetite, also known as lodestone, is easy to identify in the field because of its magnetic properties. It is usually dark or gray-black, with a metallic luster, and leaves a black streak. Crystals are rare and usually small, in the isometric habit and typically forming stubby, double-terminated octahedrons. Magnetite is quite common, as it tends to remain behind when many igneous rocks erode. Almost all rivers and streams carry some quantity of magnetite-rich "black sands" in their cracks and beneath bigger rocks. Gold panners report that there is always black sand with placer gold, but there is not always visible gold in black sands. Many prospectors save all their black sands because, in addition to magnetite, they can usually find palladium, platinum, and other rare-earth metals in the mix. California, Oregon, and northwest Washington beaches are all known for magnetite-rich sands. Lode deposits of magnetite exist in the New Jersey Highlands and the Mid-Atlantic Iron Oxide Belt of Pennsylvania.

Meteorite

This 210-pound Gibeon iron meteorite was found in Namibia, Africa.

PHOTO COURTESY OF RICE NORTHWEST MUSEUM OF ROCKS AND MINERALS

90 percent nickel to stony
Family: Extraterrestrial
Mohs: Varies by nickel-iron content
Specific gravity: Varies by nickel-iron content
Key test(s): Widmanstätten pattern
Likely locale(s): Anywhere, but deserts have slower oxidation

There are many different types of meteorites, but the main classification centers on the amount of nickel and iron present. There are iron, stony-iron, and stony meteorites. Most meteorites are actually stony, or primarily stone, but they have enough iron to at least attract a magnet on a string, so the magnet test is still warranted. Nickel is also a key component, measuring up to 7 percent, and many metal detectors can indicate the presence of nickel. Meteorites leave no streak, so a streak test that turns reddish brown is most likely hematite; a blackish-gray streak is magnetite. When sliced with a saw and etched in strong acid, iron meteorites reveal a characteristic X-shaped structure known as a Widmanstätten pattern. The best place to scout for meteorites is in a known "strewn field" surrounding an observed fall. Desert environments also are productive, because the limited moisture ensures that the iron-rich rocks don't rust and fall apart.

Molybdenite

These stunning molybdenite crystals are from Lake Chelan, Washington.

PHOTO COURTESY OF RICE NORTHWEST MUSEUM OF ROCKS AND MINERALS

Molybdenum sulfide, MoS_2
Family: Metal sulfides
Mohs: 1–1½
Specific gravity: 4.6–4.7
Key test(s): Gray to green streak; soft, greasy
Likely locale(s): Interesting mineralized zones

Molybdenite is the chief ore of molybdenum. It is somewhat rare and usually occurs only as small, dull-gray lumps. It has a metallic luster that could be confused with galena; molybdenite is also dark blue-gray to lead-gray. When big enough to conduct a scratch test, molybdenite is extremely soft—a fingernail should scratch it easily. Crystals are rare, hexagonal, and tabular, but they're easily deformed. Since molybdenite is so flexible and soft, fracture is unlikely. Graphite is also lead-gray, but graphite is lighter than molybdenite. Galena is also gray and soft and has a blue-gray streak, but galena has a cubic crystal structure. Also, molybdenite may streak slightly green on porcelain streak plates but blue-gray on paper. Noted deposits include Clear Creek, Colorado; near Lake Chelan, Washington; Inyo County, California; and large crystals in Pontiac County, Quebec.

Galena

This soft, gray galena is from lead mines in Missouri.

PHOTO COURTESY OF JIM SMITH, "IT'S JUST A ROCK" COLLECTION

Lead sulfide, PbS
Family: Metal sulfides
Mohs: 2½–2¾
Specific gravity: 7.2–7.6
Key test(s): Gray; soft; stairways
Likely locale(s): Known sulfide areas

Galena is the primary ore for lead; thus specific gravity is one of the key clues for identifying galena in the field. The dull, bluish-gray color, dull luster, and cubic, stair-stepped crystal structure are other easy clues. Finally, galena is relatively soft; you should be able to scratch it with your fingernail or a piece of calcite. Galena leaves a dark gray streak, which is fairly distinctive once you've seen it. It has a brittle fracture and can look like stibnite, the ore of antimony, but stibnite is harder. Cubes are common and distinctive as well. There is a significant lead belt in southeast Missouri near Joplin, in what are known as Mississippi Valley–type deposits, and that area has supplied prized galena specimens. Other areas with significant galena deposits are in British Columbia and Illinois. Idaho's Silver Valley has supplied significant amounts of lead in the area's silver mining operations.

Cinnabar

This bright red cinnabar in Riley opalite is from southeast Oregon.

Mercury sulfide, HgS
Family: Metal sulfide
Mohs: 2½–3
Specific gravity: 8.2
Key test(s): Crimson, pink (when fresh); hard
Likely locale(s): Calderas, limestones

Being a vivid red, cinnabar is fairly easy to identify in the field. Hexagonal crystals are quite rare; instead you should look for reddish streaks and crusts in sulfide-rich zones. If there is enough of a sample for a streak test, cinnabar leaves a scarlet streak. In rich mercury mines with heavy concentrations of cinnabar, miners have actually noted liquid mercury deposits in small pools and puddles, so look for that. As anyone who played with a broken thermometer as a child can attest, mercury is liquid at room temperature, but the vapors are dangerous. Cinnabar is heavy and bright red, but it can be confused with ochre, rust, and even common red paint. Cinnabar samples tend to be heavier than ochre, and the streak is more vivid than hematite. Look for cinnabar near hydrothermal zones, such as hot springs, or in big opalized calderas. Parts of the North American Basin and Range Province have long attracted prospectors for cinnabar, such as southern California, Nevada, and southeastern Oregon. Texas also boasts significant cinnabar deposits.

Silver

Native silver from Arizona; note the pinkish tint of "horn silver" in the sliced sample, lower left.

Ag
Family: Elemental metal
Mohs: 2½
Specific gravity: 10.5
Key test(s): Silvery (when fresh); tarnishes quickly
Likely locale(s): Economic ore deposits

Silver is actually rare in its native state because it forms oxides such as argentite so readily. Isometric crystals are rare and very collectible. Silver has a characteristic shiny, metallic luster when fresh, appearing light gray or whitish-gray at times. However, silver tarnishes quickly to black, brown, or yellow. Two key tests are for hardness and specific gravity. Lead forms crystals easier and has the characteristic stair-step pattern. Platinum is heavier than silver and even more rare, but some regions are noted for platinum nuggets, so do your research. Pure silver is exceedingly rare; there is almost always gold present, forming what miners called electrum. One of the most famous silver mining regions in North America, Idaho's Silver Valley, is not noted for great silver specimens because the ore rock is a heavy, black mass of tetrahedrite, galena, and sphalerite. Classic collecting locales for good silver specimens include Bisbee, Arizona; Creede, Colorado; Timiskaming, Ontario, and Chihuahua, Mexico.

Gold

This sample of gold ore in quartz is from the Sixteen to One Mine in the mother lode country of California.

Au
Family: Elemental metal
Mohs: 2½–3
Specific gravity: 19.3
Key test(s): Weight; color
Likely locale(s): Quartz veins; black sands

At around nineteen grams per cubic centimeter, gold is the second heaviest element, but it is also soft enough to scratch with calcite. Those two keys alone would be enough if you could find a big enough specimen to test. Instead most prospectors see tiny specks of gold in their pans, where neither test is practical. Two minerals have earned the name "fool's gold" because they mimic gold's appearance: pyrite, which is brassy, harder, and smells like sulfur when crushed; and mica, which breaks easily with a knifepoint and tends to float. Gold has a characteristic metallic luster and produces a golden streak if you dare rub it on a streak plate. Raw gold is rarely, if ever, pure; it is usually alloyed with silver, which affects the color significantly. Prospectors wash gold from gravels using tools such as pans, sluices, dredges, and highbankers. The Klondike area of Alaska and California's mother lode country are most noted for good placer ground and bigger gold specimens, but most Northern and Western areas have at least some microscopic gold.

PART 3:
GEMS

Precious Opal

This spectacular precious opal is from Spencer, Idaho.

PHOTO COURTESY OF SPENCER OPAL MINES

SiO_2*nH_2O
Family: Metal silicates
Mohs: 5½–6½
Specific gravity: 2.0–2.2
Key test(s): Play of color; luster
Likely locale(s): Virgin Valley; Spencer

Precious opal is a rare form of common opal that displays a characteristic play of all colors, sometimes subtle and sometimes quite dramatic. Precious opal owes its stunning color to the way its silica spheres are stacked and packed, with distances impacting diffracted light and causing the color interplay. Precious opal is relatively soft, has no crystal structure or cleavage, and is usually found in veins and nodules. The streak is white. Precious opal comes in a variety of forms: standard precious opal, displaying all colors of the rainbow, usually after replacing wood; fire opal, usually red or orange, derived from thin seams between lava flows; black opal, very dark in color but with a nice play of colors; and white opal. Two famed locales in North America are at Virgin Valley, Nevada, where the precious opal replaces wood and sometimes bone, and Spencer, Idaho, where the opal occurs in thin seams. Both are fee-dig operations.

Turquoise

This polished turquoise nodule is from Bisbee, Arizona.

PHOTO COURTESY OF RICE NORTHWEST MUSEUM OF ROCKS AND MINERALS

$CuAl_6(PO_4)_4(OH)_8*4-5H_2O$
Mohs: 5–7
Specific gravity: 3.5–4.0
Key test(s): Blue; waxy luster
Likely locale(s): Veins, seam fillings

Turquoise is practically synonymous with Southwest Native American jewelry, so it is sometimes strange to see it without a polished silver setting around it. Sadly, because demand is so high, some turquoise is dyed or impregnated to enhance its color and bring out more of the coveted blue. Natural turquoise rarely forms crystals but instead is found as nuggets and filled-in fractures in the host rock. Turquoise frequently forms veins and nodules as a secondary replacement mineral. Turquoise has a waxy luster and a faint, bluish-white streak; the powder is soluble in hydrochloric acid. Under long-wave ultraviolet light, turquoise may fluoresce. Black limonite veining is common. Minerals confused with turquoise include chrysocolla, which is much softer, and variscite, which is usually greener and softer. Look for turquoise in known copper-producing areas where phosphates are also common. Famous locales include the area around Cerillos, New Mexico, one of the oldest known producers; a pair of high-quality mines in Arizona; and a long belt of turquoise across Nevada.

Sunstone

Here are three forms of sunstone: Oregon sunstone in basalt matrix (upper left); piles of unpolished (upper right) and polished (bottom, middle) sunstones.

(Ca,Na)((Al, Sl)$_2$Si$_2$O$_8$)
Mohs: 6–6½
Specific gravity: 2.64–2.66
Key test(s): Shape; color; hardness
Likely locale(s): Basalt

Sunstones are a form of plagioclase feldspar so are yet another incarnation of a sodium calcium aluminum silicate. They are usually yellow, but in Oregon the inclusion of tiny copper crystals provides a unique "Schiller" effect that can be deep red and makes for an interesting effect as the specimen rotates. Sunstones form in basalt vesicles and are usually very fractured and irregular. The euhedral crystals have a vitreous luster, a white streak, and cleave in two directions. They don't look anything like the orbs and blebs of opal- or agate-filled basalt vesicles and should be scratched by quartz. Sunstones take a nice polish. One of the most famous sunstone locales is at Rabbit Hills, near Plush, Oregon. The Bureau of Land Management has set aside a free area there, and there are fee-dig operations nearby, such as the famed Dust Devil Mine.

Garnet

These samples of raw and polished garnet are from Emerald Creek, Idaho and the largest specimen is from Brazil.

PHOTO COURTESY OF GENE HARDGROVE, WWW.ROCKARTGEMS.COM

Almandine, $X_3Y_2(SiO_4)_3$
Family: Metal silicates
Mohs: 6½–7½
Specific gravity: 3.6–4.3
Key test(s): Cubic crystals; hardness
Likely locale(s): Schist; black sands

Increasing metamorphism ⟶

Chlorite	Biotite	Garnet	Staurolite	Kyanite	Sillimanite

The garnet family is a complicated group of six main varieties, with varying amounts of iron, magnesium, aluminum, and calcium represented in the X_3Y_2 formula above. Taken as a family, garnet crystals show a vitreous luster, have no streak to speak of, and do not cleave. The range of colors found in garnets is vast, but red and purple are very common. Garnet is harder than apatite, does not fluoresce like zircon, and has higher specific gravity than tourmaline, plus it is usually associated with schist. That rule isn't hard and fast, however, as some gem garnets are associated with granitic pegmatites. One of the most famous star garnet locales is the fee-dig operation run by the USDA Forest Service at Emerald Creek, Idaho.

Jade

This translucent apple-green jade comes from Wyoming.

Manganese silicate, MnSiO$_3$
Mohs: 6½–7 for jadeite; 5½–6 for nephrite
Specific gravity: 2.9–3.1
Key test(s): Can't be scratched by a knife; botyroidal
Likely locale(s): Mafic rocks

There are two main varieties of jade. Both are amazingly strong due to interlocking crystals that form nearly unbreakable bonds. First there is classic jadeite—a sodium- and aluminum-rich pyroxene found mostly in Burma and favored by Chinese rulers for centuries. Jadeite is not quite as hard as quartz. Second there is nephrite jade—a type of amphibole, also famous in China but more commonly associated with North American deposits in Wyoming, British Columbia, Washington State, and California. Nephrite is slightly softer than jadeite and usually occurs only as white or shades of green. A high-quality steel knife blade cannot scratch either jadeite or nephrite. Jadeite, which can be purple, blue, lavender, pink, or vivid green, is the more highly prized, but that isn't a hard-and-fast rule, as rare, "mutton-fat" white nephrite jade commands fabulous prices. One famous North American locale for jadeite, on the Motagua River in Guatemala, was rediscovered thanks to landslides from Hurricane Mitch in 1998. The source of this highly prized blue-green jade, known as Olmec blue jade, had been lost for centuries prior to the hurricane.

GEMS

Beryl

This large green beryl crystal is an emerald from the Hiddenite Emerald District in North Carolina.

PHOTO COURTESY OF ED SPEER; WWW.NORTHCAROLINAEMERALDS.INFO

Beryllium aluminum silicate, $Be_3Al_2Si_6O_{18}$
Mohs: 7½–8
Specific gravity: 2.7–2.9
Key test(s): Red, green, or blue; hexagonal form
Likely locale(s): Granite pegmatites

Like corundum, beryl comes in a variety of colors, each with its own name. Bright green beryl is known as emerald, while blue beryl is known as aquamarine. When red or pink, beryl is called morganite, but there is a striking red variety from Utah known as "scarlet emerald." Other varieties include a golden beryl known for its striking yellow color and heliodor, which is greenish-yellow. Goshenite is the name for colorless varieties of beryl. In all forms, beryl has a vitreous luster and a colorless streak. Crystals are hexagonal, sometimes in striking, perfectly six-sided prisms. Cleavage is not a good test, but hardness is, as beryl is harder than quartz. Common beryl is a constituent of many pegmatites in North America, including the Kaniksu Batholith of northeast Washington and northern Idaho. Gem-quality beryl typically occurs in vuggy cavities and pockets in pegmatite accompanied by mica, quartz, and feldspar. North Carolina's Hiddenite Emerald District, which has several fee-dig operations, is the preemininent North American collecting locale for emeralds.

Topaz

These twinned topaz crystals are from Topaz Mountain, Utah.

Aluminum silicate, $Al_2SiO_4(F,OH)_2$
Mohs: 8
Specific gravity: 3.4–3.6
Key test(s): Pink (when fresh); hard
Likely locale(s): Pegmatites; rhyolite cavities

Topaz is an interesting mineral that doesn't seem to get much respect for its natural state; it is commonly irradiated or coated to produce more pleasing colors. Typically topaz is clear when pure, but impurities can cause it to appear white, light gray, or even pink or yellow. Laboratory treatments result in more stunning shades of blue and orange. Topaz has a glassy to vitreous luster and forms stubby, prismatic crystals in the orthorhombic crystal system, with excellent terminations. The best test in the field is its hardness—topaz will scratch quartz. Cleavage is excellent in one direction, and striations are common lengthwise on crystal faces, but topaz is often found in massive lumps. Topaz forms at high temperatures among silica-rich igneous rocks such as rhyolite or in cavities within granite pegmatites. The most prominent collecting locale in North America is at Topaz Mountain, Utah, where topaz crystals formed in vugs and cavities within rhyolite.

Corundum

These samples are Montana sapphires (right) and red ruby in the rough.

RUBY PHOTO COURTESY OF JIM SMITH, "IT'S JUST A ROCK" COLLECTION

Aluminum oxide, Al_2O_3
Mohs: 9
Specific gravity: 3.9–4.1
Key test(s): Only diamond is harder
Likely locale(s): Pegmatites; alluvial deposits

North American corundum comes in three main varieties: Emery is black or dark gray and is historically used for abrasives. Sapphire can be blue, yellow, purple, or green. Pink and red are the two colors reserved for rubies. The only difference is the amount and kind of impurities, typically iron, titanium, or chromium. Each variety has a hardness of 9 on the Mohs scale, so all forms of corundum are scratched only by diamond. Luster ranges from vitreous all the way to adamantine, especially in true gem-quality rubies and sapphires. The hexagonal crystal habit is not a good field indicator, mostly because it is rare to find crystals big enough to inspect. Certain pegmatites can contain corundum, as can certain gneisses and hornfels. There are large deposits of emery near Peekskill, New York. There are two well-known areas in North America to search for rubies and sapphires, which usually occur together in some ratio. Gem-quality rubies and sapphires headline at fee-dig operations in Franklin, North Carolina, and at Spokane Bar, near Helena, Montana.

Diamond

This two-carat yellow diamond was recovered in 2010 from Arkansas.

PHOTO COURTESY OF GLENN WORTHINGTON

Carbon, C
Mohs: 10
Specific gravity: 3.5
Key test(s): Hardness; luster (adamantine)
Likely locale(s): In kimberlite pipes

Diamonds are a serious challenge to identify in the field for several reasons. First, any small chip of agate or quartz or even calcite may look a little like a small diamond. Second, diamonds tend to travel. There are countless reports of gold miners finding diamonds in their sluice boxes at cleanup, probably thousands of miles away from their source. Good luck getting a streak—diamonds will scratch any streak plate. Color is completely variable, with specimens coming in clear, yellow, pink, blue, purple, brown, and even black. The hardness test is the best indicator: At 10 on the Mohs scale, pure, strongly crystallized diamond scratches everything. You can pick up a piece of topaz or corundum at a gem shop and have it on hand for your test. Clear agate and clear gem-quality quartz such as Herkimer "diamonds" look similar to actual diamonds but are much softer. Diamonds are related to kimberlite pipes, unique intrusives found in very few places worldwide. The best area to search for diamonds in North America is the fee-dig Crater of Diamonds State Park in Murfreesboro, Arkansas.

Glossary

Alluvium: Dirt, usually. Stream and river deposits of sand, mud, rock, and other material. Sometimes sorted if laid down in deep water; otherwise can be unsorted if deposited during floods, earthquakes, etc. If glaciers were involved, the term till is used.

Basement: The "lowest" and oldest rocks around; usually metamorphic and frequently dating to the Precambrian or Paleozoic era. Although considered the basement rocks, they are usually less prone to erosion and can actually make up mountain ranges and stunning cliffs. "Basement" refers to their placement at the bottom of a stratigraphic table.

Batholith: General term that refers to extremely large masses of coarse intrusive rock such as granite, quartz monzonite, granodiorite, or diorite that extend over a large surface area. Anything over 100 km^2 is considered a batholith. For perspective, the Idaho Batholith covers about 15,000 square miles. Other significant batholiths include the Sierra Nevada range, Pikes Peak in Colorado, South Mountain in Nova Scotia, and Stone Mountain in Georgia.

Bleb: A round or oval cavity, air bubble, hole, or vesicle; usually in basalt and sometimes filled with opal, agate, or chalcedony.

Clasts: Catchall term for the clay, silt, sand, gravel, cobbles, and boulders that make up nonchemical sedimentary rocks and other breccias. The size of the clasts then determines the name of the rock.

Density: Since it isn't enough to measure outright weight of two different samples, we need to define the weight per an agreed unit of mass. By weighing the sample and then dunking it in water and measuring the volume of water displaced, we get the density measured in grams per cubic centimeter. The general term "heft" refers to a field test for how dense a hand specimen feels.

Era: The four main geologic eras, from oldest to youngest, are the Precambrian, Paleozoic, Mesozoic, and Cenozoic.

Float: Describes the difference between rock samples hammered from an outcrop, and thus with a known origin, and samples that exist as cobbles or boulders and not attached to bedrock. Prospectors are able to trace float to its source outcrop.

Formation: This is a key term to understand in field geology. Geologists assign formation names to mappable, recognizable rock assemblages and also note a "type" locale that defines the rest of the unit. Formations can be lumped together into groups or even supergroups, such as the Belt

Supergroup of northern Idaho, Montana, British Columbia, Washington, and Wyoming. To become a formation, a group of similar rocks must be big enough to be worth the bother, must share some key similarity, and must be traceable across the surface. Some formations are divided into members.

Hydrothermal vein: Good spot to investigate for interesting minerals. These hot, chemical-rich solutions, usually quartz, can either find an existing crack in country rock or create their own. If they cool slowly enough, hydrothermal veins can create large crystals prized by collectors.

Intrusion: Catchall term for the various granites, diorites, and related rocks that bulldoze their way through the earth's crust but never reach the surface. Cooling in place quickly results in fine-grained material; cooling slowly gives the individual elements more time to build up into larger crystals. Understanding intrusions is a key to understanding geology, but nobody has ever witnessed an intrusion, so all of our assumptions are based on field evidence and educated guesswork.

Lode: A prized zone of rich, extended mineralization that usually ensures a successful mining operation. The term is reserved for larger vein networks that cover significant ground, such as the Mother Lode gold mining district of California's Sierra Nevada.

Luster: Term for the visual appearance of a mineral's lighted surface. The way minerals reflect light can be helpful for identification, but there is no scientific measurement for luster. These brief definitions should help guide you.

Luster Term	Description
Adamantine	Shiny
Chatoyant	Cat's-eye
Earthy	Very dull
Greasy	Slick
Pearly	Like pearls
Resinous	Like amber
Schiller	Platy inclusions
Silky	Shines directionally
Vitreous	Glasslike
Waxy	Dull

Native metal: Metals in their purest form are not significantly combined with oxides, sulfides, carbonates, or silicates and are thus native. Gold, silver, copper, platinum, and mercury are examples of metals sometimes found in their native condition.

Ore: The term used to describe a viable mineral deposit that is worth mining. Usually refers to a metals-based mineral that must be milled.

Outcrop: A cliff, ledge, or other visible clue to the rock formations below.

Reaction series: Refers to the behavior of a cooling magma where some minerals form at high temperature but as the magma cools further, the early minerals dissolve and reform as a new, different mineral. These conditions are observable in a laboratory setting.

Replacement deposit: Describes a particular type of ore deposit where hot, circulating solutions first dissolve a mineral to form a cavity and then fill the void with a new material.

Stratigraphic column: Stratigraphy is the science and study of sedimentary rock outcrops. Stratigraphers create elegant stratigraphic columns that pictorially represent the measured or inferred relations of rock outcrops, measuring the thickness of groups, formations, and members. Metamorphic and igneous rocks occasionally show up in stratigraphic columns, but stratigraphy is primarily a tool to understand sedimentary rocks.

Streak: Refers to the color of the powdered mineral dust left behind when a mineral is scraped across a streak plate. The color of this fine rock powder is a more true reflection of the mineral than visual appearance.

Texture: Describes a rock's grain size, crystal size, if the grains are uniform or variable, if the grains are rounded or angular, and if there is any evidence of orientation to the grains.

Zeolite: Common aluminosilicates formed in volcanic rocks, such as basalt, where alkaline groundwaters circulate at low temperature and create a ringed "molecular sieve" structure. Very useful for filtering contaminants and absorbing odors.

Index